Macrophyte and pH Buffering Updates to the Klamath River Water-Quality Model Upstream of Keno Dam, Oregon

By Annett B. Sullivan and Stewart A. Rounds, U.S. Geological Survey; Jessica R. Asbill-Case, Bureau of Reclamation; Michael L. Deas, Watercourse Engineering, Inc.

Prepared in cooperation with the Bureau of Reclamation

Scientific Investigations Report 2013–5016

U.S. Department of the Interior
U.S. Geological Survey

U.S. Department of the Interior
KEN SALAZAR, Secretary

U.S. Geological Survey
Suzette M. Kimball, Acting Director

U.S. Geological Survey, Reston, Virginia: 2013

For more information on the USGS—the Federal source for science about the Earth, its natural and living resources, natural hazards, and the environment, visit http://www.usgs.gov or call 1–888–ASK–USGS.

For an overview of USGS information products, including maps, imagery, and publications,
visit http://www.usgs.gov/pubprod

To order this and other USGS information products, visit http://store.usgs.gov

Suggested citation:
Sullivan, A.B., Rounds, S.A., Asbill-Case, J.R., and Deas, M.L., 2013, Macrophyte and pH buffering updates to the Klamath River water-quality model upstream of Keno Dam, Oregon: U.S. Geological Survey Scientific Investigations Report 2013-5016, 52 p.

Contents

Figures

Tables

Conversion Factors, Datums, and Abbreviations

Conversion Factors

Inch/Pound to SI

Multiply	By	To obtain
foot (ft)	0.3048	meter (m)
mile (mi)	1.609	kilometer (km)
cubic inches (in.^{3})	0.01639	liter (L)

SI to Inch/Pound

Multiply	By	To obtain
micron (μm)	0.00003937	inch (in)
millimeter (mm)	0.03937	inch (in)
meter (m)	3.281	foot (ft)
square meter (m^2)	1,550.0	square inch (in^2)
gram (g)	0.03527	ounce, avoirdupois (oz)
kilogram (kg)	2.205	pound avoirdupois (lb)
liter (L)	33.82	ounce, fluid (fl oz)
liter (L)	0.2642	gallon (gal)
milliliter (mL)	0.03382	ounce, fluid (fl oz)
liter (L)	61.02	cubic inch (in^3)
cubic meter (m^3)	35.31	cubic foot (ft^3)
cubic meter (m^3)	264.2	gallon (gal)
milligram per liter (mg/L)	1.0	parts per million (ppm)
Watt (W)	3.41	British thermal unit per hour (BTU/h)

Temperature in degrees Celsius (°C) may be converted to degrees Fahrenheit (°F) as follows:

°F=(1.8×°C)+32.

Concentrations of chemical constituents in water are given either in milligrams per liter (mg/L), which is approximately equivalent to parts per million (ppm), or micrograms per liter (μg/L), which is approximately equivalent to parts per billion (ppb).

Datums

Elevation refers to distance above the vertical datum. A local vertical datum (UKLVD) is used, established by the Bureau of Reclamation. For the purpose of this report, the conversion is UKLVD − 1.78 ft = NGVD29.

Horizontal coordinate information is referenced to the North American Datum of 1983 (NAD 83).

Abbreviations

K_a	acid dissociation constant
<	less than
≤	less than or equal to
DOC	dissolved organic carbon
>	greater than
≥	greater than or equal to
pK_a	acid dissociation constant
MAE	mean absolute error
POC	particulate organic carbon
POM	particulate organic matter
SRP	soluble reactive phosphorus
TMDL	total maximum daily load
USGS	U.S. Geological Survey

Macrophyte and pH Buffering Updates to the Klamath River Water-Quality Model Upstream of Keno Dam, Oregon

By Annett B. Sullivan[1], Stewart A. Rounds[1], Jessica R. Asbill-Case[2], and Michael L. Deas[3]

Abstract

A hydrodynamic, water temperature, and water-quality model of the Link River to Keno Dam reach of the upper Klamath River was updated to account for macrophytes and enhanced pH buffering from dissolved organic matter, ammonia, and orthophosphorus. Macrophytes had been observed in this reach by field personnel, so macrophyte field data were collected in summer and fall (June–October) 2011 to provide a dataset to guide the inclusion of macrophytes in the model. Three types of macrophytes were most common: pondweed (*Potamogeton* species), coontail (*Ceratophyllum demersum*), and common waterweed (*Elodea canadensis*). Pondweed was found throughout the Link River to Keno Dam reach in early summer with densities declining by mid-summer and fall. Coontail and common waterweed were more common in the lower reach near Keno Dam and were at highest density in summer. All species were most dense in shallow water (less than 2 meters deep) near shore. The highest estimated dry weight biomass for any sample during the study was 202 grams per square meter for coontail in August. Guided by field results, three macrophyte groups were incorporated into the CE-QUAL-W2 model for calendar years 2006–09. The CE-QUAL-W2 model code was adjusted to allow the user to initialize macrophyte populations spatially across the model grid.

The default CE-QUAL-W2 model includes pH buffering by carbonates, but does not include pH buffering by organic matter, ammonia, or orthophosphorus. These three constituents, especially dissolved organic matter, are present in the upper Klamath River at concentrations that provide substantial pH buffering capacity. In this study, CE-QUAL-W2 was updated to include this enhanced buffering capacity in the simulation of pH. Acid dissociation constants for ammonium and phosphoric acid were taken from the literature. For dissolved organic matter, the number of organic acid groups and each group's acid dissociation constant (K_a) and site density (moles of sites per mole of carbon) were derived by fitting a theoretical buffering response to measured upper Klamath River alkalinity titration curves. The organic matter buffering in the Klamath River was modeled with two monoprotic organic acids: carboxylic acids with a mean pK_a of 5.584 and site density of 0.1925, and phenolic organic acids with a mean pK_a of 9.594 and site density of 0.6466. Total inorganic carbon concentrations in the model boundary inputs were recalculated based on the new buffering equations. CE-QUAL-W2 was also adjusted to allow the simulation of nonconservative alkalinity caused by nitrification, denitrification, photosynthesis, and respiration. The Klamath River model was recalibrated after the macrophyte and pH buffering updates producing improved predictions for pH, dissolved oxygen, and particulate carbon.

Introduction

The Klamath River flows about 255 mi (410 km) from the outlet of Upper Klamath Lake to the Pacific Ocean through southern Oregon and northern California. The first 1-mile reach downstream of Upper Klamath Lake (from Link River Dam to near sampling site 11507501; fig. 1) is named Link River; the Klamath River proper begins at the mouth of Link River. Stage in the next 20 miles is controlled by Keno Dam (fig. 1). Water quality in the Link River to Keno Dam reach has been designated as "water quality limited" for exceeding ammonia toxicity and dissolved oxygen criteria year-round, and pH and chlorophyll *a* criteria in summer (Oregon Department of Environmental Quality, 2007). A Total Maximum Daily Load (TMDL) plan for this reach of the Klamath River (Oregon Department of Environmental Quality, 2010) was approved by the U.S. Environmental Protection Agency in May 2012. Water temperature allocations have also been established for point and nonpoint sources in this reach, due to the water temperature TMDL below Keno Dam.

[1]U.S. Geological Survey

[2]Bureau of Reclamation

[3]Watercourse Engineering, Inc.

these samples. However, dissolved organic matter in the Klamath River system is likely to have multiple sources, such as wetlands, algae, or woody debris from historical and current mill operations. Further investigation into the sources of organic matter in the Klamath River system would help to better understand the variability of that material and its effect on pH buffering.

A second optimization was designed to test whether two monoprotic acids with no distribution of associated acidities could be used to fit the organic-matter portion of the experimental titration curves. After all, with the rather small pK_a standard deviations found in the first optimization, the fitted Gaussian distribution of pK_a values was almost identical to just two monoprotic acids. In this second optimization, only four organic-matter parameters were fitted using the theoretical titration curve: two site densities and two pK_a values. The results were quite similar to and slightly better than the results from the first optimization, with a mean absolute volume error of about 10.1 counts per titration (800 counts to 1 mL of acid volume). The mean absolute volume error from the first optimization (6 parameters) was about 10.2 counts per titration.

Erring on the side of simplicity, it was decided to model the pH buffering of dissolved organic matter in the Klamath River using just two monoprotic acids and the parameters listed in table 4 for optimization #2. Although this is a simple approach, the resulting theoretical titration curves do a remarkably good job of fitting the experimental titration curves. Three examples from the 24 fitted datasets are shown in figures 5A, B, and C, in which the measured titration curves (points) are compared against four different theoretical titration curves. The simplest theoretical titration curve accounts only for the chemistry of carbonic acid and water, in which hydroxide, carbonate, and bicarbonate are the only entities that can be titrated. Successive theoretical curves then add the effects of ammonia, phosphates, and dissolved organic matter. In most of the Klamath River samples having a high pH, the carbonate-only theoretical titration curve failed to capture the shape of the measured titration curve, particularly for the higher-pH part of the curve. Adding in the effects of ammonia often helped when ammonia concentrations were high (close to or higher than 1.0 mg/L as nitrogen) and the sample pH was above 8.5. Phosphates rarely had a visible effect on the theoretical titration curve for the range of phosphate concentrations in these samples. Dissolved organic matter had a far more important and substantive effect on the theoretical titration curve for these samples and was instrumental in capturing the shape of the measured titration curve. Clearly, for these Klamath River samples, dissolved organic matter appears to be an important constituent that can buffer the pH of the river when algal blooms consume carbonic acid and force the pH to levels of 8 and above.

Table 4. Dissolved organic-matter acid dissociation constants and site densities for the Link River to Keno Dam reach of the Klamath River, Oregon, resulting from two best-fit optimizations of 24 measured alkalinity titration curves.

[**Abbreviations:** pK_a, acid dissociation constant; std. dev., standard deviations; –, not applicable]

Acid group	Model parameter	Optimization number 1 (2 acid distributions)	Optimization number 2 (2 monoprotic acids)
1 (Carboxylics)	Site density	0.1751	0.1925
	pK_a mean	5.502	5.584
	pK_a std. dev.	0.047	–
2 (Phenolics)	Site density	0.5828	0.6466
	pK_a mean	9.738	9.594
	pK_a std. dev.	0.050	–

Although the nature of the dissolved organic matter is likely to change over the course of a season and as a function of its various sources, representing its buffering with a pair of monoprotic acids in the model calculations appears to be a good first step in capturing that process.

Recalculation of Boundary Conditions for Total Inorganic Carbon

With the inclusion of organic matter, orthophosphorus, and ammonia into the internal CE-QUAL-W2 calculations of pH, it is also necessary to adjust the concentration of total inorganic carbon in the model input files. Prior to the enhanced buffering updates, the concentration of total inorganic carbon for the inflows into the model was calculated from field measurements of pH, sample titrations to measure alkalinity, and an assumption that the alkalinity is predominantly due to bicarbonate, carbonate, and hydroxide. When that is the case, the alkalinity equation was simplified:

$$\text{Alk} \approx \text{Alk}_{carb} = [\text{HCO}_3^-] + 2[\text{CO}_3^{2-}] + [\text{OH}^-] - [\text{H}^+] \quad (40)$$

and the determination of the total inorganic carbon concentration was relatively straightforward once the values of the acid dissociation constants for carbonic acid are known:

$$C_T = \frac{\{\text{H}^+\}^2 + K_1'\{\text{H}^+\} + K_1'K_2'}{K_1'\{\text{H}^+\} + 2K_1'K_2'}\left(\text{Alk} - \frac{K_w'}{\{\text{H}^+\}} + \frac{\{\text{H}^+\}}{\gamma_{\text{H}^+}}\right) \quad (41)$$

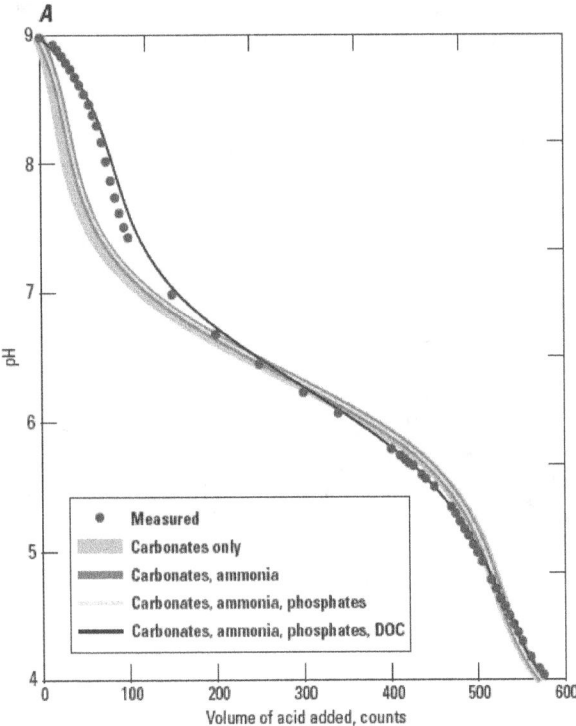

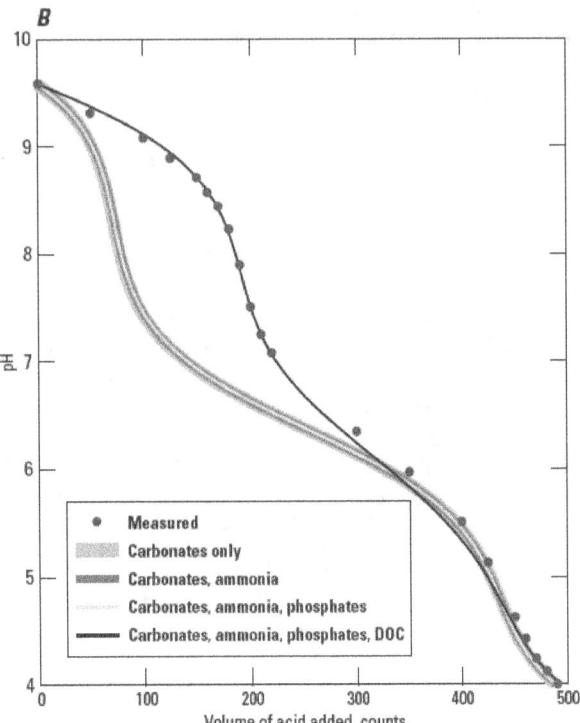

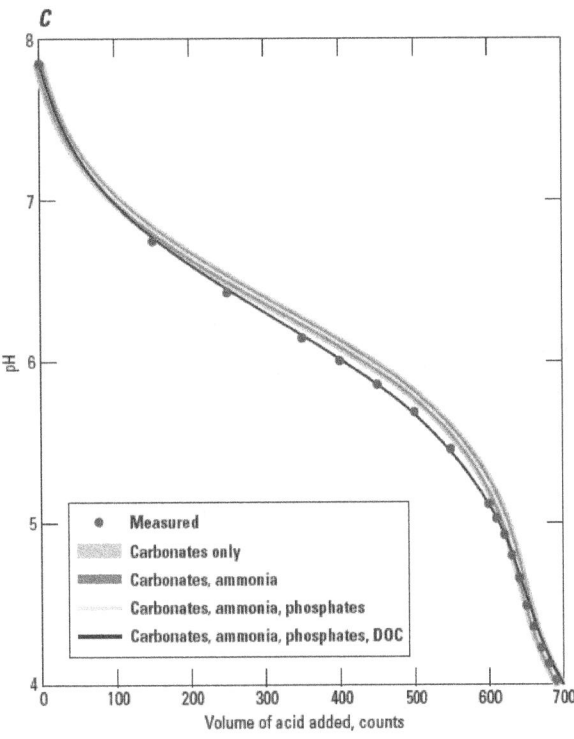

Figure 5.   Example titration curves of four theoretical titration curves and measured data from the Link River to Keno Dam reach of the Klamath River, Oregon. DOC, dissolved organic carbon. (*A*) In this sample at Miller Island July 17, 2007, alkalinity was 52.8 mg/L as $CaCO_3$, ammonia concentration was 1.1 mg/L as N, phosphate concentration was 0.171 mg/L, and DOC concentration was 11.1 mg/L. Ammonia had a visible effect on the theoretical titration curve, but organic matter was important in capturing the shape of the measured curve. (*B*) In this sample at Link River August 14, 2007, alkalinity was 44.2 mg/L as $CaCO_3$, ammonia concentration was 0.068 mg/L as N, phosphate concentration was 0.104 mg/L, and DOC concentration was 11.4 mg/L. Ammonia had little effect on the shape of the curve because the ammonia concentration was low, but organic matter was important to the shape of the titration curve. (*C*) In this sample at Keno August 14, 2007, alkalinity was 65.2 mg/L as $CaCO_3$, ammonia concentration was 1.01 mg/L as N, phosphate concentration was 0.165 mg/L, and the DOC concentration was 12.5 mg/L. Because the sample pH was less than 8.0, ammonia and organic matter had little effect on the high end of the titration curve, but organic matter had a slight effect on the lower part of the curve.

Activating the enhanced pH buffering routines in CE-QUAL-W2 means that, to the extent that such enhanced buffering is important compared to that provided by bicarbonate and carbonate, the boundary condition total inorganic carbon concentration must be computed by taking into account the contributions of Alk_{am}, Alk_P, and Alk_{OM} if those are activated in the model. Taking all three into account, the equation for C_T becomes:

$$C_T = \frac{\{H^+\}^2 + K_1'\{H^+\} + K_1'K_2'}{K_1'\{H^+\} + 2K_1'K_2'} \left(Alk - \frac{K_w'}{\{H^+\}} + \frac{\{H^+\}}{\gamma_{H^+}} - N_T \frac{K_{am}'}{\{H^+\} + K_{am}'} \right.$$
$$\left. - P_T \left(\frac{K_{P1}'K_{P2}'\{H^+\} + 2K_{P1}'K_{P2}'K_{P3}' - \{H^+\}^3}{\{H^+\}^3 + K_{P1}'\{H^+\}^2 + K_{P1}'K_{P2}'\{H^+\} + K_{P1}'K_{P2}'K_{P3}'} \right) - A_T \sum_{i=1}^{n_a} \delta_i \left(\frac{1}{1+10^{(pK_{a_i}-pH)}} - \frac{1}{1+10^{(pK_{a_i}-4\,5)}} \right) \right) \qquad (42)$$

If neglecting activity corrections in a freshwater system, a slightly simpler equation results:

$$C_T = \frac{[H^+]^2 + K_1[H^+] + K_1K_2}{K_1[H^+] + 2K_1K_2} \left(Alk - \frac{K_w}{[H^+]} + [H^+] - N_T \frac{K_{am}}{[H^+] + K_{am}} \right.$$
$$\left. - P_T \left(\frac{K_{P1}K_{P2}[H^+] + 2K_{P1}K_{P2}K_{P3} - [H^+]^3}{[H^+]^3 + K_{P1}[H^+]^2 + K_{P1}K_{P2}[H^+] + K_{P1}K_{P2}K_{P3}} \right) - A_T \sum_{i=1}^{n_a} \delta_i \left(\frac{1}{1+10^{(pK_{a_i}-pH)}} - \frac{1}{1+10^{(pK_{a_i}-4\,5)}} \right) \right) \qquad (43)$$

where
- C_T is the total inorganic carbon concentration (in moles per liter; multiply result by 12,011 to obtain milligrams per liter),
- K_1 is the first acid dissociation constant for carbonic acid (table 3),
- K_2 is the second acid dissociation constant for carbonic acid (table 3),
- K_w is the acid dissociation constant for water (table 3),
- K_{am} is the acid dissociation constant for ammonium (table 3),
- K_{P1} is the first acid dissociation constant for phosphoric acid (table 3),
- K_{P2} is the second acid dissociation constant for phosphoric acid (table 3),
- K_{P3} is the third acid dissociation constant for phosphoric acid (table 3),
- Alk is the alkalinity in equivalents per liter [(Alk in milligrams per liter $CaCO_3$)/50,044],
- N_T is the concentration of ammonia and ammonium in moles per liter [(ammonia+ammonium in mg/L)/14,006.74],
- P_T is the concentration of SRP in moles per liter [(SRP in milligrams per liter)/30,973.762],
- n_a is the number of monoprotic acid groups, or the number of distributions of those groups, used to represent organic matter,
- A_T is the total organic matter concentration in moles of carbon per liter. Depending on model options either [(DOC in milligrams per liter)/12,011] or [(DOC+POC in milligrams per liter)/12,011],
- δ_i is the site density for acid group i, in moles of sites per mole of carbon,
- K_{a_i} is the acid dissociation constant of the ith acid group for organic matter,
- pK_{a_i} is the negative logarithm, base 10, of K_{a_i} (for distribution option (DIST), use 0.5, 1.0, 1.5, ..., 13.0, 13.5),
- pH is the pH of the water, and
- $[H^+]$ is the hydrogen ion concentration, defined as 10^{-pH}.

For organic matter, if the user chooses to represent the organic acids with a distribution of pK values, the details of the final distribution of pK values is needed. The ph_buffering. opt model output file from a test run with the organic matter inputs provides the overall site densities for a range of pK values to use with organic matter. The site density is the moles of acid/base groups per mole of carbon. For the distribution (DIST) option, 27 pK values from 0.5 to 13.5, at 0.5-unit increments, are used to characterize the final distribution and need to be accounted for in the calculation of C_T.

The calculation of total inorganic carbon concentrations from these equations can be used if the acid dissociation constants and site densities for organic matter have been determined, as in the upper Klamath River, or in different systems if the buffering provided by organic matter was smaller and could be ignored. It is best when simulating organic-matter buffering to analyze a series of alkalinity titration curves in advance of the modeling to estimate values for the organic matter site densities and pK_a values. Otherwise, using these values as calibration parameters would require that the boundary conditions for C_T be recomputed whenever the organic-matter parameter values that affect pH buffering are modified.

Activity Corrections

The PH_CO2 subroutine in CE-QUAL-W2 (appendix B) includes activity correction calculations for most of the inorganic constituents included in the computations. Most of that code was left unchanged in this version, with the exception that some additional constituents not included in the default implementation were given activity corrections. Ionic strength was estimated in the default code based on simulated values of total dissolved solids or salinity. A form of the extended Debye-Hückel equation was used in the default code to estimate activity coefficients for bicarbonate, carbonate, hydroxide, and the hydrogen ion (H^+). In this version, additional activity coefficients were computed for ammonia, ammonium, and the various phosphate species based on the existing equations and a set of ion size and charge factors tabulated for the Debye-Hückel equation by Stumm and Morgan (1996).

Nonconservative Alkalinity

Previous versions of CE-QUAL-W2 assumed that alkalinity is largely unaffected by chemical and biological reactions and responds only to changes in inputs, transport, and mixing. While alkalinity is largely a conservative quantity, it is not entirely conservative. Reactions such as ammonia nitrification and nitrate denitrification as well as photosynthesis and respiration can have small effects on alkalinity. A summary of these effects was provided by

Stumm and Morgan (1996) in their aquatic chemistry text (see table 4.5, page 173). The following reactions have effects on alkalinity:

- Utilization of ammonium during photosynthesis results in a decrease in alkalinity: 14 equivalents of alkalinity for every 16 moles of ammonium used. (*Note: This stoichiometry was applied despite the fact that this equation is not balanced in H.*)

$$106CO_2 + 16NH_4^+ + HPO_4^{2-} + 108H_2O$$
$$\rightarrow C_{106}H_{263}O_{110}N_{16}P + 107O_2 + 14H^+$$

- Utilization of nitrate during photosynthesis results in an increase in alkalinity: 18 equivalents of alkalinity for every 16 moles of nitrate used.

$$106CO_2 + 16NO_3^- + HPO_4^{2-} + 122H_2O + 18H^+$$
$$\rightarrow C_{106}H_{263}O_{110}N_{16}P + 138O_2$$

- Production of ammonium during respiration results in an increase in alkalinity: 14 equivalents of alkalinity for every 16 moles of ammonium produced. (*Note: This stoichiometry was applied despite the fact that this equation is not balanced in H.*)

$$C_{106}H_{263}O_{110}N_{16}P + 107O_2 + 14H^+$$
$$\rightarrow 106CO_2 + 16NH_4^+ + HPO_4^{2-} + 108H_2O$$

- Nitrification of ammonium results in a decrease in alkalinity: 2 equivalents of alkalinity for every 1 mole of ammonium consumed.

$$NH_4^+ + 2O_2 \rightarrow NO_3^- + H_2O + 2H^+$$

- Denitrification of nitrate (to nitrogen gas) results in an increase in alkalinity: 1 equivalent of alkalinity per 1 mole of nitrate consumed.

$$5CH_2O + 4NO_3^- + 4H^+ \rightarrow 5CO_2 + 2N_2 + 7H_2O$$

As a result of changes made in this version of CE-QUAL-W2, the user is allowed to specify whether alkalinity is to be simulated as a conservative or a nonconservative quantity (see appendix B).

Summary of Updated Model Calibration

After macrophytes and enhanced pH buffering were incorporated into the upper Klamath River model, the model was recalibrated by adjusting selected parameter values to obtain a good match between measured data and model output. Most parameters were unchanged from the original model calibration and remain as documented by Sullivan and others (2011). Those parameters that were adjusted are shown in table 5. The new macrophyte parameters, as used in the upper Klamath River model, are shown in table 2 and the enhanced buffering parameters are shown in table 6.

During macrophyte calibration, the macrophyte field data were used to provide insight into the general patterns of density, spatial and temporal distributions of biomass, and macrophyte species characteristics. The model was not calibrated directly to the macrophyte field data because the macrophyte field data were collected in a different year (2011) than the model years (2006–09). Examples of the modeled distribution of the three macrophyte groups through the model domain on July 2 and August 17 for model year 2007 are shown in figure 6. Although most constituents in CE-QUAL-W2 are laterally averaged across the channel, macrophytes are modeled in a quasi three-dimensional mode so that macrophytes can be modeled at different depths from bank to bank across a model segment (Berger, 2000; Berger and Wells, 2008; Cole and Wells, 2008). As in the field data, the modeled pondweed density was highest in early summer and diminished by August, whereas coontail and common waterweed were simulated with a higher density in late summer and were more common in the downstream reaches near Keno Dam.

Macrophytes affect the cycles of oxygen, nutrients, organic matter, and pH. In the updated model, the total annual production of dissolved oxygen by macrophytes is approximately the same order of magnitude as that by algae (fig. 7). Oxygen consumption by respiration is higher in macrophytes compared to algae, due in part to the fact that the macrophyte oxygen respiration parameter was adjusted to account for the large population of snails within the macrophytes. Dissolved oxygen production and consumption by macrophytes begins earlier in the year than oxygen production and consumption by algae, largely because the pondweed macrophytes are early season species, whereas the largest algal blooms occur in late June and early July.

Table 5. Parameter values changed as a result of calibration updates to the water-quality model for the Link River to Keno Dam reach of the Klamath River, Oregon.

[**Abbreviations:** C, degrees Celsius; g, gram; 1/day, per day]

Parameter	Original value	Updated value	Description
AG, blue-green	3.09	2.3	Maximum algal growth rate for blue-green algae, 1/day
AT1, blue-green	12	13	Lower temperature parameter for rising rate function for blue-green algae, °C
AT3, blue-green	35	30	Lower temperature parameter for falling rate function for blue-green algae, °C
CO2R	1.2	0.1	Sediment carbon dioxide release rate, fraction of sediment oxygen demand

Table 6. Enhanced pH buffering input parameters in the water-quality model for the Link River to Keno Dam reach of the Klamath River, Oregon.

[**Abbreviations:** pK_a, acid dissociation constant; OM, organic matter; POM, particulate matter]

Parameter	Value	Description
PHBUFC	ON	Option to use enhanced buffering routines
NCALKC	ON	Option to use nonconservative alkalinity algorithms
NH4BUFC	ON	Specifies whether ammonia/ammonium is included in pH buffering
PO4BUFC	ON	Specifies whether phosphoric acid is included in pH buffering
OMBUFC	ON	Specifies whether organic matter is included in pH buffering
OMTYPE	MONO	MONO specifies discrete pK_a values
NAG	2	Number of organic acid/base groups to model
POMBUFC	OFF	Specifies whether POM is included in OM buffering
SDEN1	0.1925	Site density for organic group 1, in moles of acid/base sites per mole of carbon in OM
SDEN2	0.6466	Site density for organic group 2, in moles of acid/base sites per mole of carbon in OM
PK1	5.584	The pK_a values (negative log10 of the acid dissociation constant)
PK2	9.594	The pK_a values (negative log10 of the acid dissociation constant)

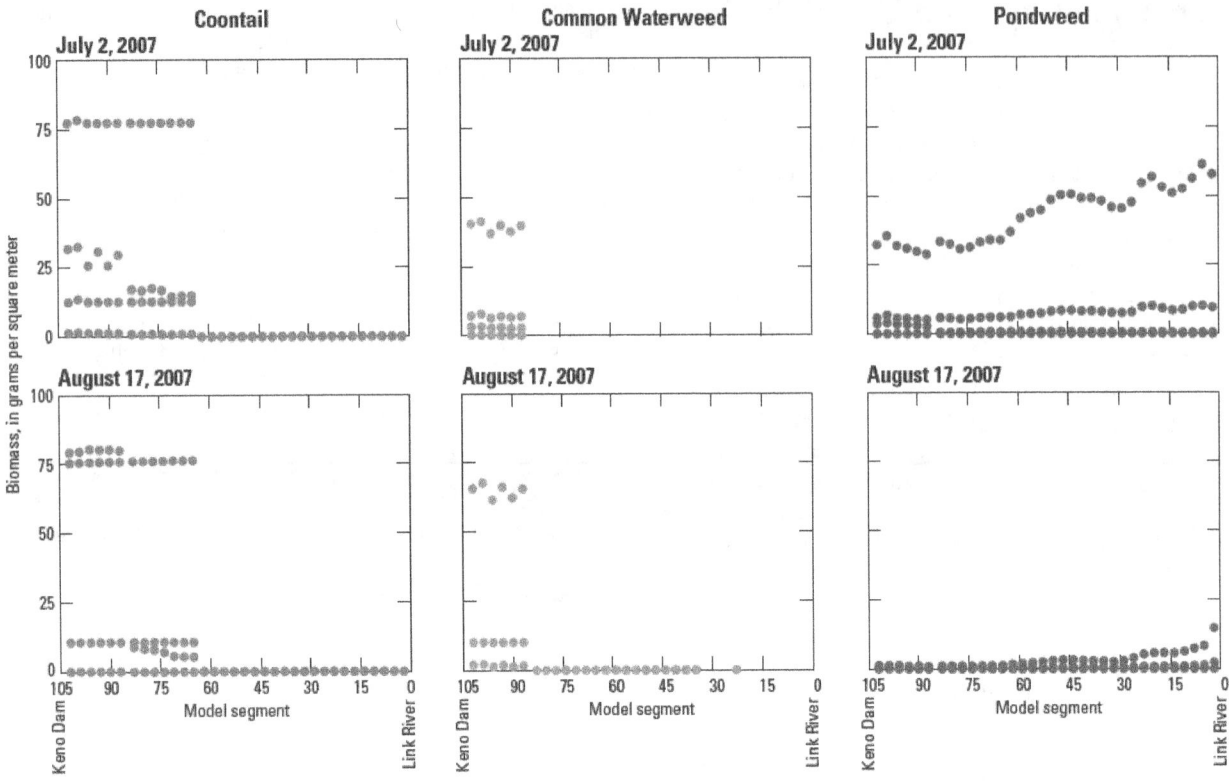

Figure 6.   Modeled macrophyte biomass along the Link River to Keno Dam reach of the Klamath River, Oregon, on two selected dates. CE-QUAL-W2 provides macrophyte output at different depths across each segment cross section.

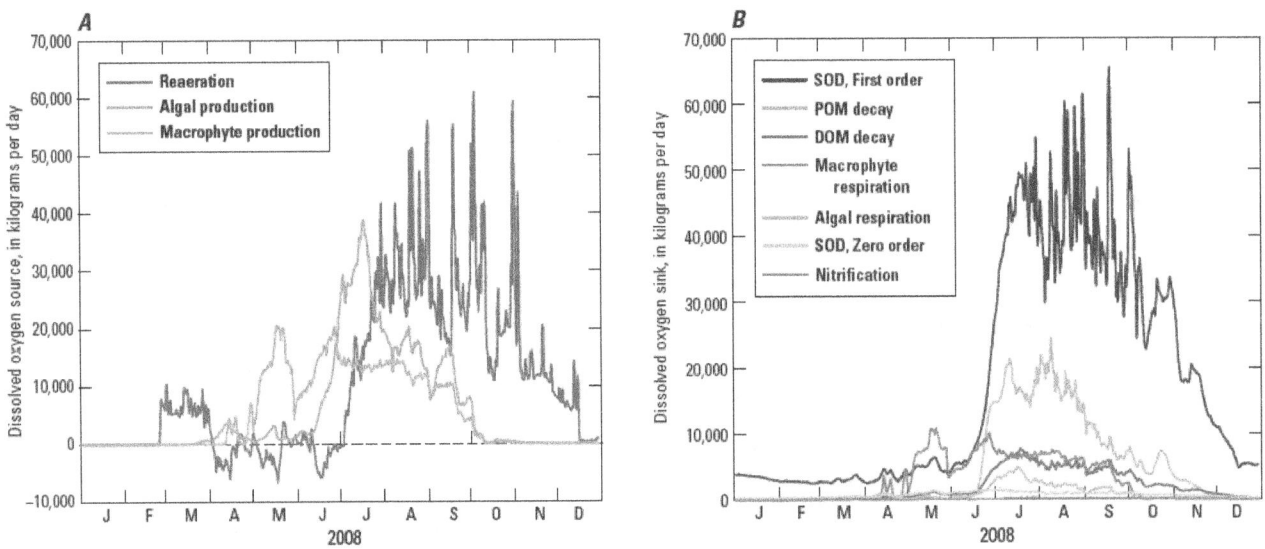

Figure 7.   Modeled daily average dissolved-oxygen production (*A*) sources and consumption (*B*) sinks in the Link River to Keno Dam reach of the Klamath River, Oregon, in 2008. DOM, dissolved organic matter; POM, particulate organic matter; SOD, sediment oxygen demand.

pH buffering by ammonia, phosphates, and dissolved organic matter was enabled in the recalibrated model, but buffering by particulate organic matter was disabled (table 6). Particulate organic matter, much of it derived from dying algae, may have some acid/base properties, but being enclosed in particles likely inhibits its reactivity with the bulk river water. Algae and macrophyte material may also have acid/base properties, such as cell walls with carboxyl groups (Chojnacka, 2010), but this buffering was also assumed to be small relative to buffering by dissolved organic matter. Nonconservative alkalinity was turned on, and two monoprotic acids were used to simulate the dissolved organic matter buffering—one to represent carboxylic acids and one to represent phenolic and amine groups together.

The enhanced pH buffering algorithms had a notable effect on modeling pH in summer and fall. Without the enhanced buffering, the modeled pH was generally lower and subject to larger daily variability compared to measured pH (fig. 8). The addition of macrophytes also improved pH modeling, especially in the lower part of the reach near Keno Dam where macrophytes were more abundant (fig. 9). With both macrophytes and enhanced pH buffering implemented, the model was able to simulate pH well with mean absolute errors (MAE) of ≤0.34 for years 2006–09 (fig. 10; table 7). The inclusion and calibration of pH in the upper Klamath River model will now allow model scenarios to examine the effect of management and other system changes on pH, which is a water-quality limited constituent in this reach.

The updates to the model affected the error statistics of other constituents modestly. The most improved error statistics were for dissolved oxygen, with an average MAE improvement of 0.1 mg/L, and particulate carbon, with an average MAE improvement of 0.2 mg/L. Average error statistics for other constituents were essentially the same compared to the original USGS model (Sullivan and others 2011). In addition to the model performance metrics included in table 7, measured and modeled pH and dissolved oxygen time series are shown in figures 10A–D and 11A–D, respectively, for calendar years 2006–09. Algae, nutrients, organic matter, and bottom sediment are shown in figures 12A–D for the calendar years 2006–09. Modeled temperature and specific conductance are not shown here; those plots are available in the original USGS Klamath River model report (Sullivan and others, 2011).

When interpreting measured/model comparison plots and error statistics, remember that model output and measured data are at the same time and location but not at the exact same scale. For example, measurements from a continuous monitor probe represent a small volume near the probe tip. Model output is from an entire model cell that might have dimensions of 1,000 ft × 500 ft × 2 ft near-surface. Measured data and model output should show the same large-scale temporal and spatial patterns, but some mismatch at smaller scales can be expected as a result of these different scales.

The USGS upper Klamath River model has already been used to analyze water-quality effects of system changes in several preliminary model scenarios (Sullivan and others, 2011, 2012). The predictive improvements provided by the updated model will benefit future analyses of water-quality scenarios and help improve the understanding of water-quality processes in this reach of the Klamath River.

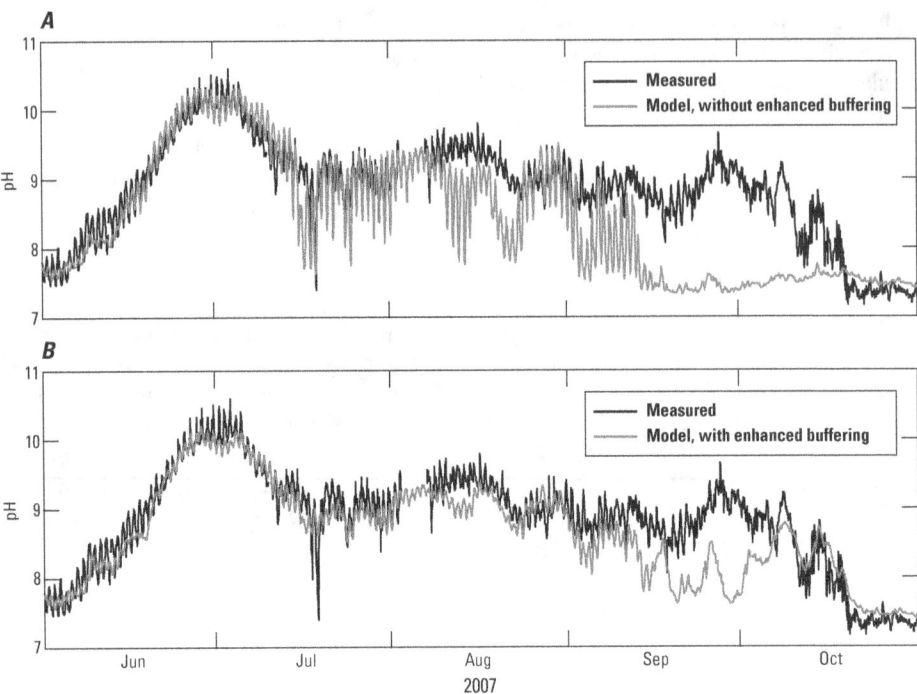

Figure 8.   Modeled and measured pH at the Klamath River, Oregon, Railroad Bridge site in summer and fall 2007, (*A*) without and (*B*) with the new enhanced CE-QUAL-W2 pH buffering. To illustrate just the effect of the enhanced pH buffering, neither of the model runs shown includes macrophytes.

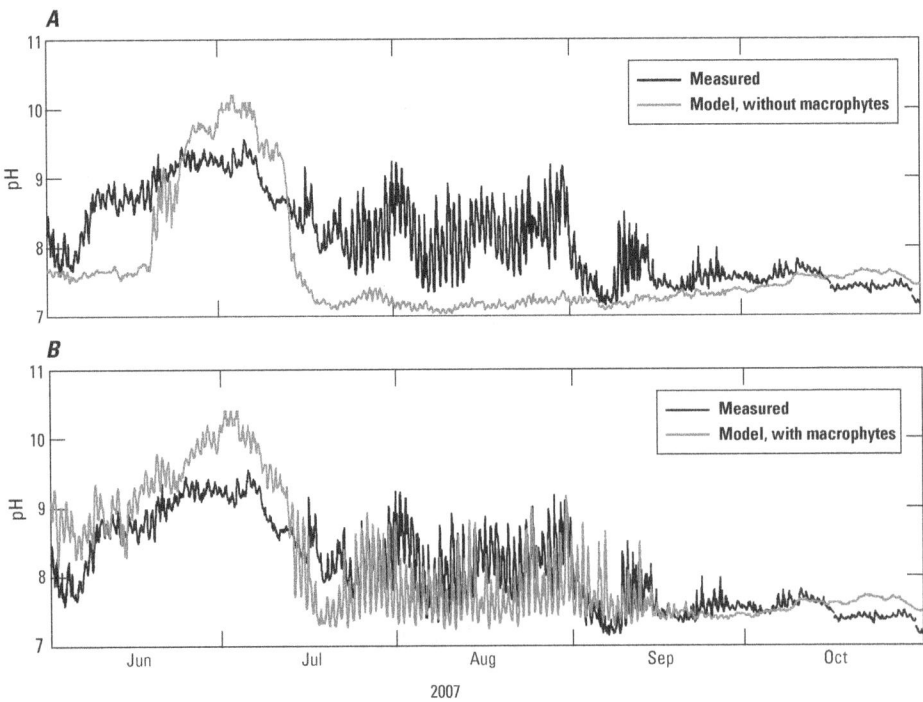

Figure 9.   Modeled and measured pH at the Klamath River, Oregon, Keno site in summer and fall 2007, (*A*) without macrophytes and (*B*) with macrophytes. To illustrate just the effect of macrophytes, neither of the model runs shown includes enhanced pH buffering.

Table 7. Goodness-of-fit statistics averaged over all calibration sites for the updated model of the Link River to Keno Dam reach of the Klamath River, Oregon.

[**Abbreviations:** C. degrees Celsius; <, less than; μm³/mL, cubic micrometers per milliliter; mg/L, milligrams per liter; rl, reporting level; ×, times]

Constituent	Unit	Data type	Range of measured values	Year	Mean error	Mean absolute error	Root mean square error
Water temperature	°C	hourly	0–28	2006	-0.03	0.55	0.70
				2007	0.06	0.63	0.78
				2008	0.17	0.63	0.80
				2009	-0.17	0.58	0.72
Dissolved oxygen	mg/L	hourly	0–20	2006	-0.02	0.98	1.37
				2007	0.70	1.23	1.70
				2008	-0.18	0.95	1.36
				2009	-0.65	1.27	1.71
pH		hourly	7–11	2006	-0.14	0.26	0.36
				2007	-0.04	0.30	0.39
				2008	-0.02	0.25	0.33
				2009	-0.20	0.34	0.46
Ammonia	mg/L	grab	<rl–1.7	2007	0.00	0.12	0.16
				2008	-0.11	0.22	0.28
Nitrate	mg/L	grab	<rl–0.8	2007	0.01	0.03	0.03
				2008	0.01	0.03	0.03
Particulate nitrogen	mg/L	grab	0.06–4.0	2007	-0.21	0.29	0.49
				2008	-0.17	0.25	0.39
Total nitrogen	mg/L	grab	0.6–5.9	2007	0.03	0.40	0.59
				2008	-0.08	0.42	0.52
Orthophosphorus	mg/L	grab	0.01–0.27	2007	0.01	0.03	0.04
				2008	0.00	0.02	0.03
Total phosphorus	mg/L	grab	0.06–0.50	2007	0.01	0.05	0.07
				2008	0.00	0.04	0.05
Blue-green algae	μm³/mL	grab	0–113×10⁶	2007	-4.7×10⁶	7.6×10⁶	14.9×10⁶
				2008	-2.2×10⁶	3.4×10⁶	5.8×10⁶
Particulate carbon	mg/L	grab	0.5–18	2007	0.07	1.09	1.72
				2008	0.41	1.31	1.89
Dissolved organic carbon	mg/L	grab	5–14	2007	0.37	0.72	1.12
				2008	-0.39	0.71	0.88

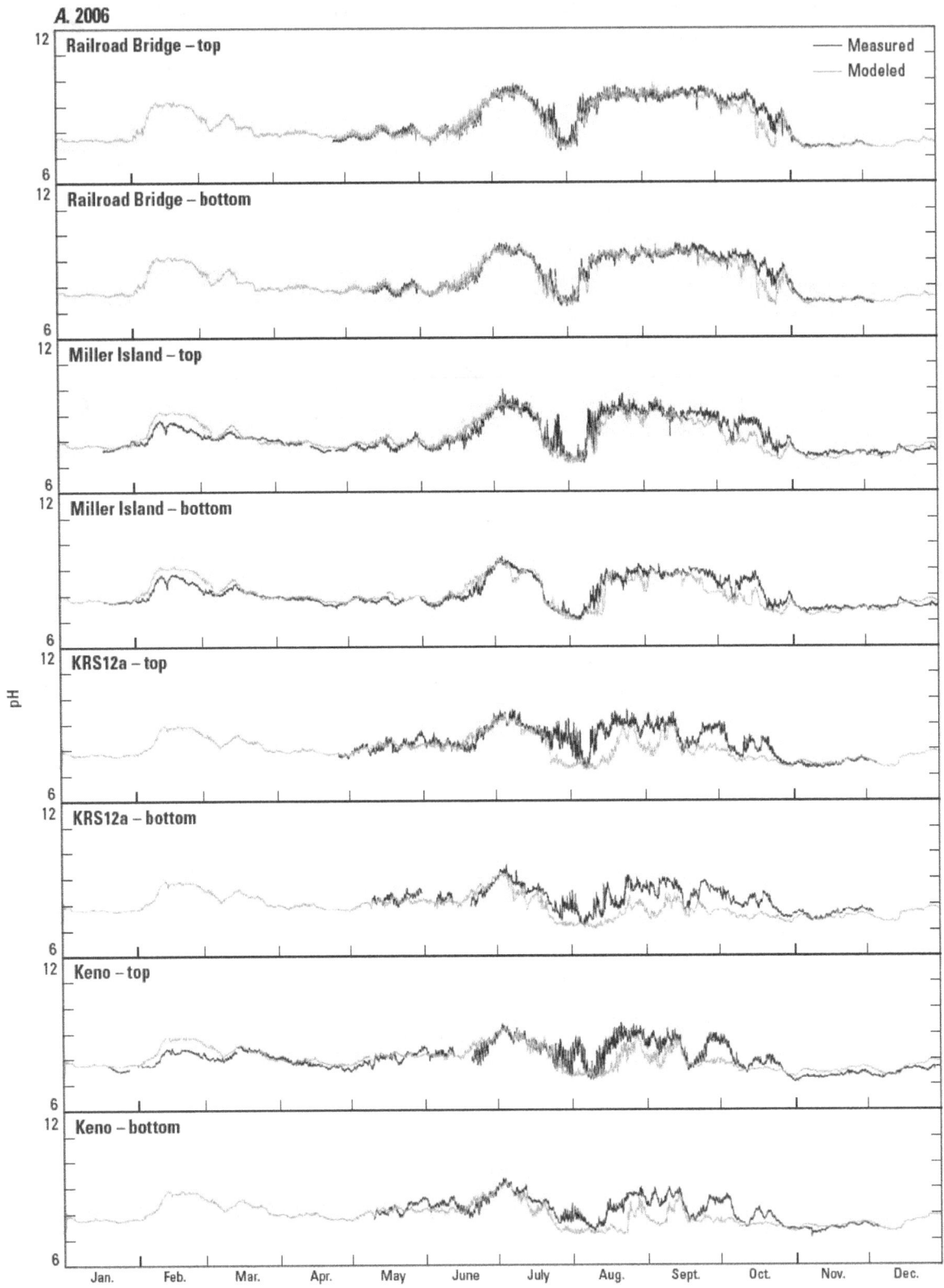

Figure 10.   Measured and modeled hourly pH during calendar years (*A*) 2006, (*B*) 2007, (*C*) 2008, and (*D*) 2009 for sites in the Link River to Keno Dam reach of the Klamath River, Oregon. Top results were from 1 meter below the river surface; bottom results were from 1 meter above the channel bottom.

B. 2007

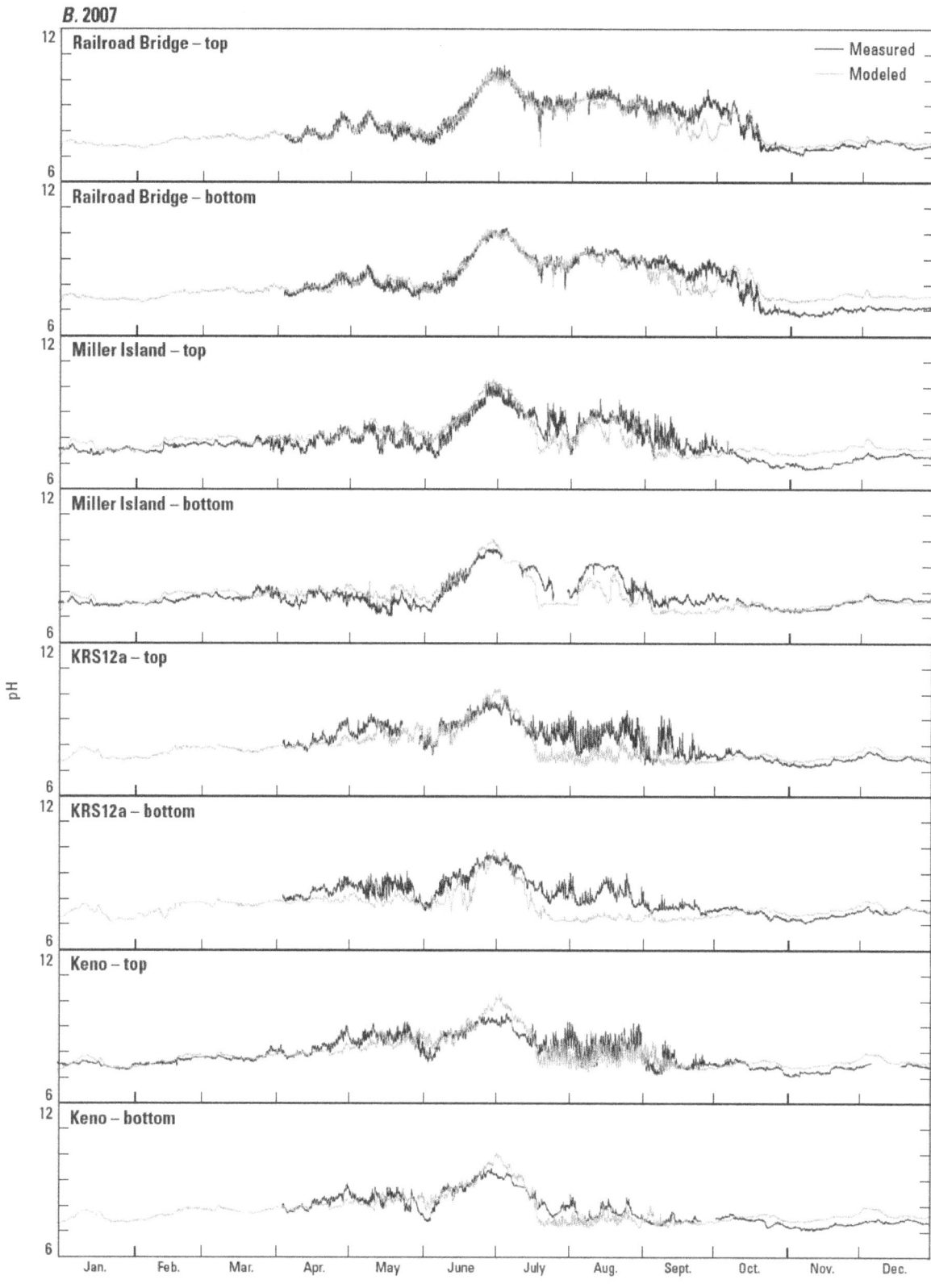

Figure 10.—Continued

C. 2008

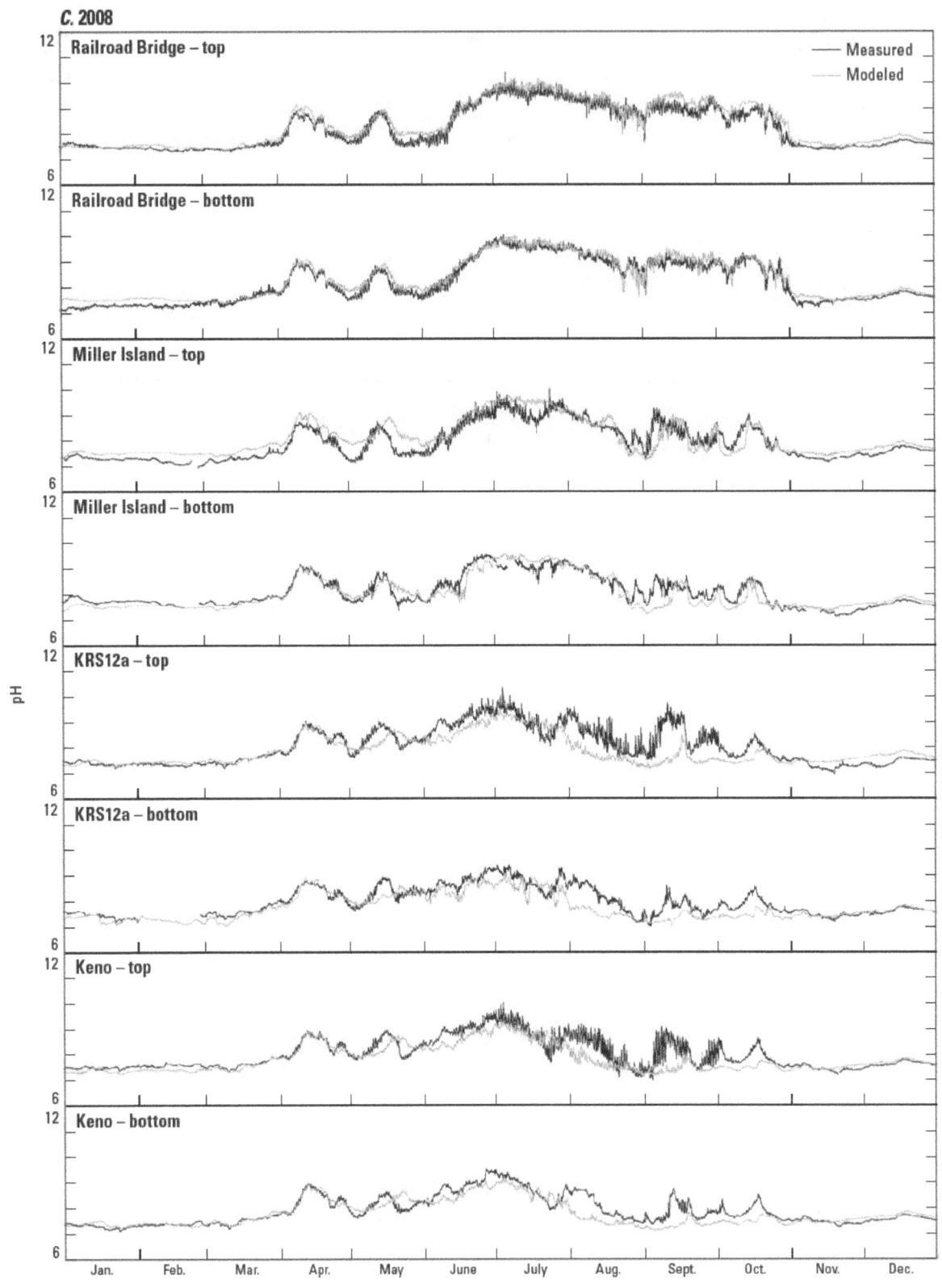

Figure 10.—Continued

D. 2009

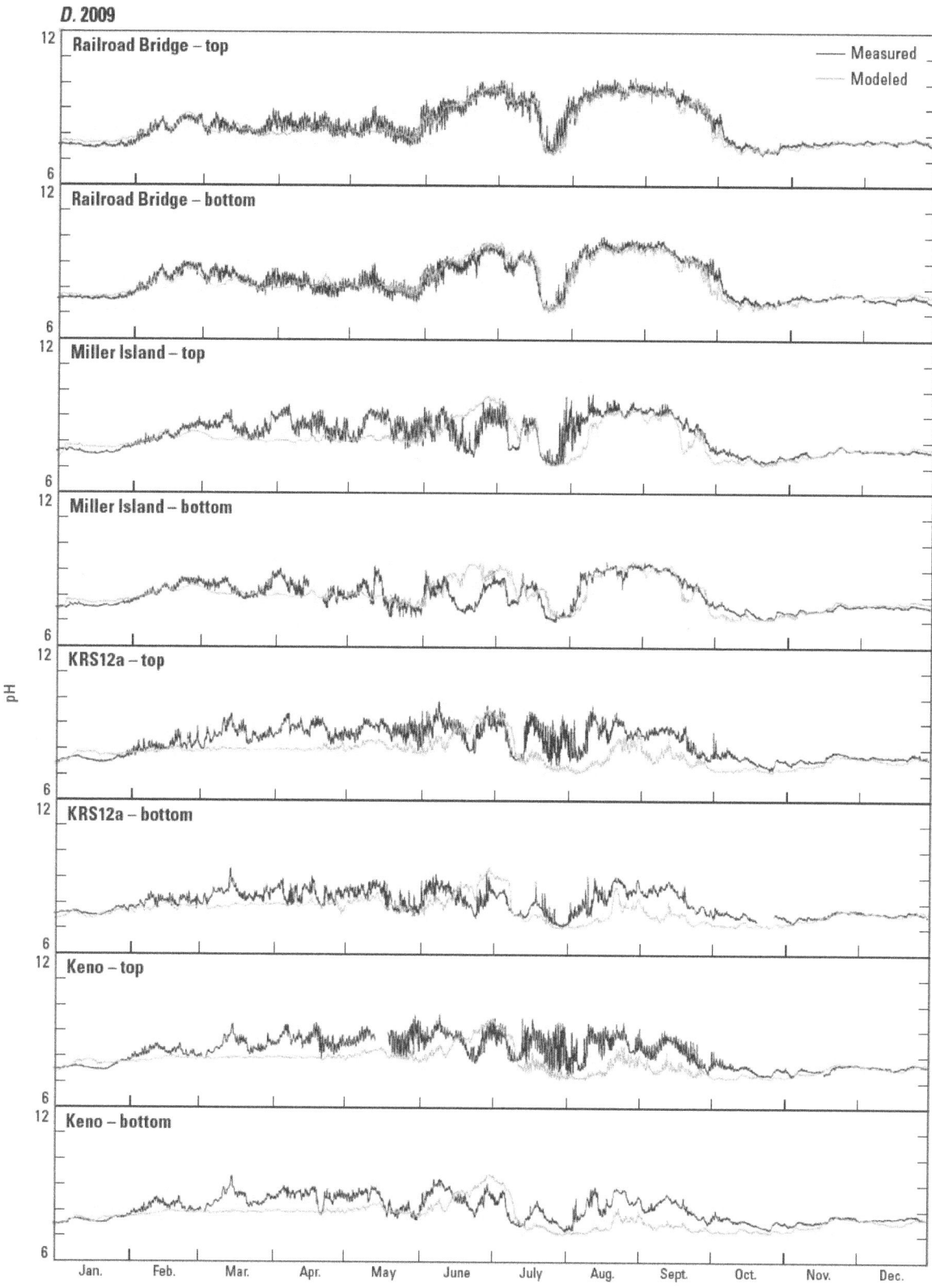

Figure 10.—Continued

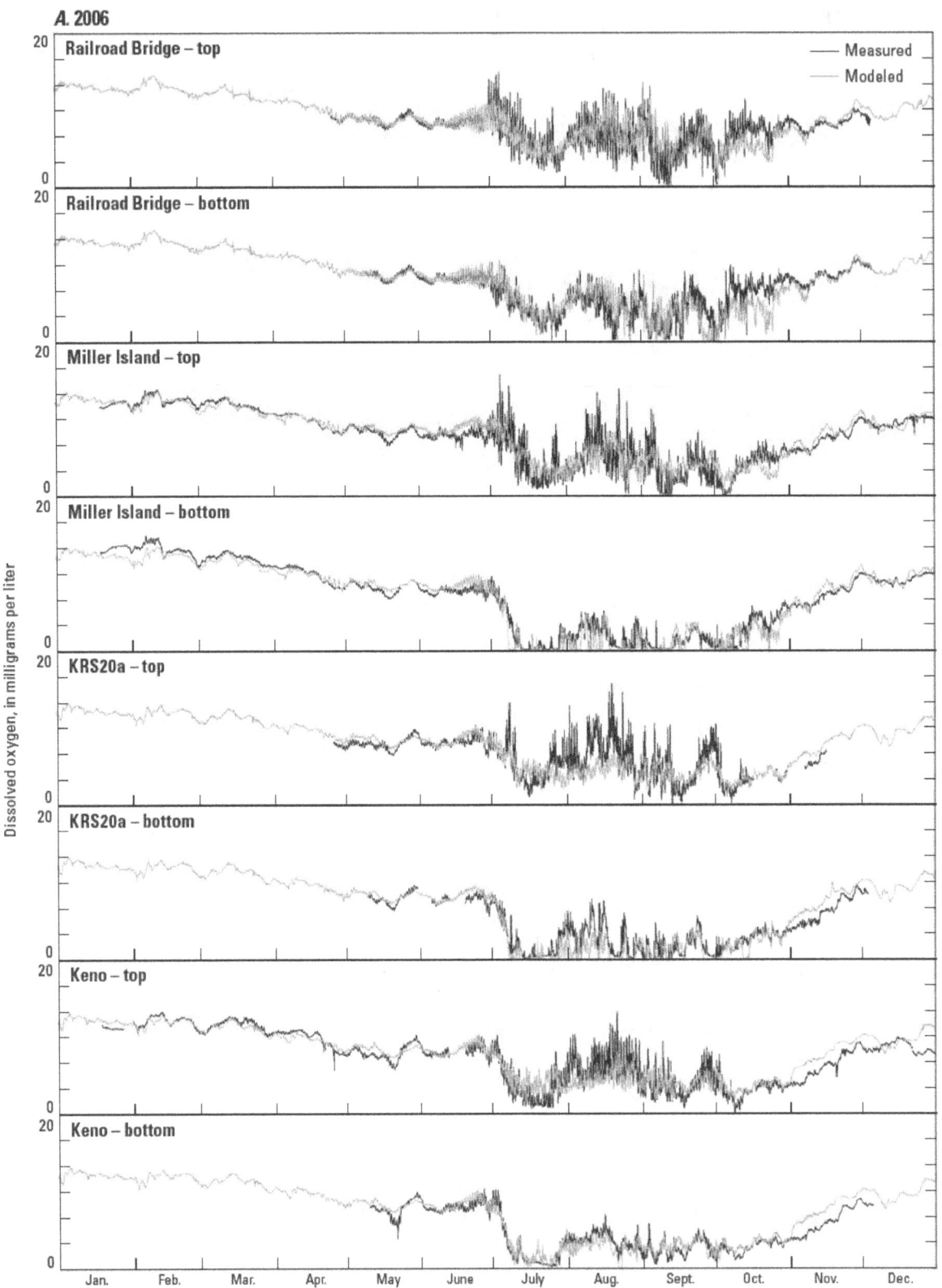

Figure 11. Measured and modeled hourly dissolved-oxygen concentrations during calendar years (*A*) 2006, (*B*) 2007, (*C*) 2008, and (*D*) 2009 for sites in the Link River to Keno Dam reach of the Klamath River, Oregon. Top results were from 1 meter below the river surface; bottom results were from 1 meter above the channel bottom.

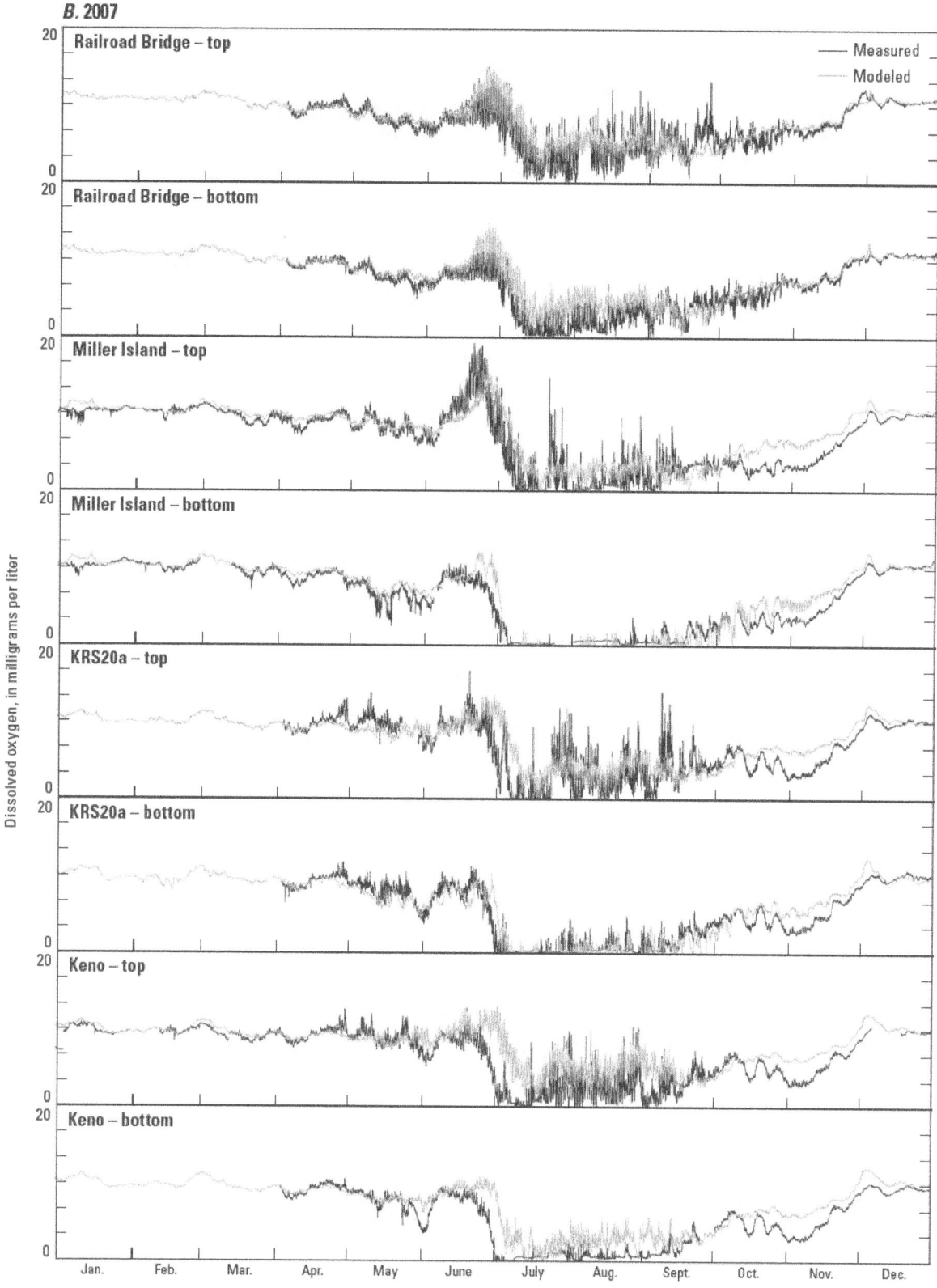

Figure 11.—Continued

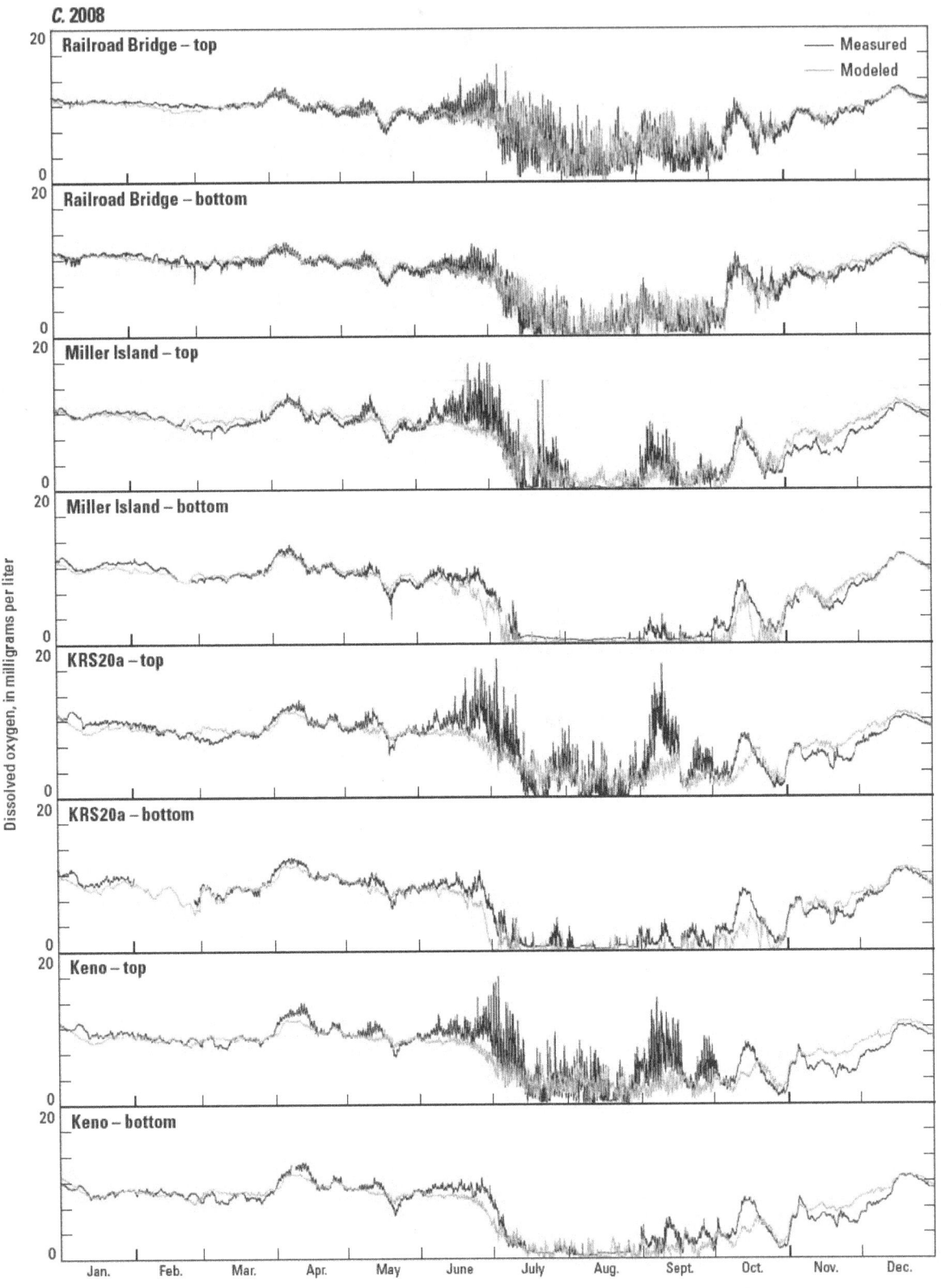

Figure 11.—Continued

D. 2009

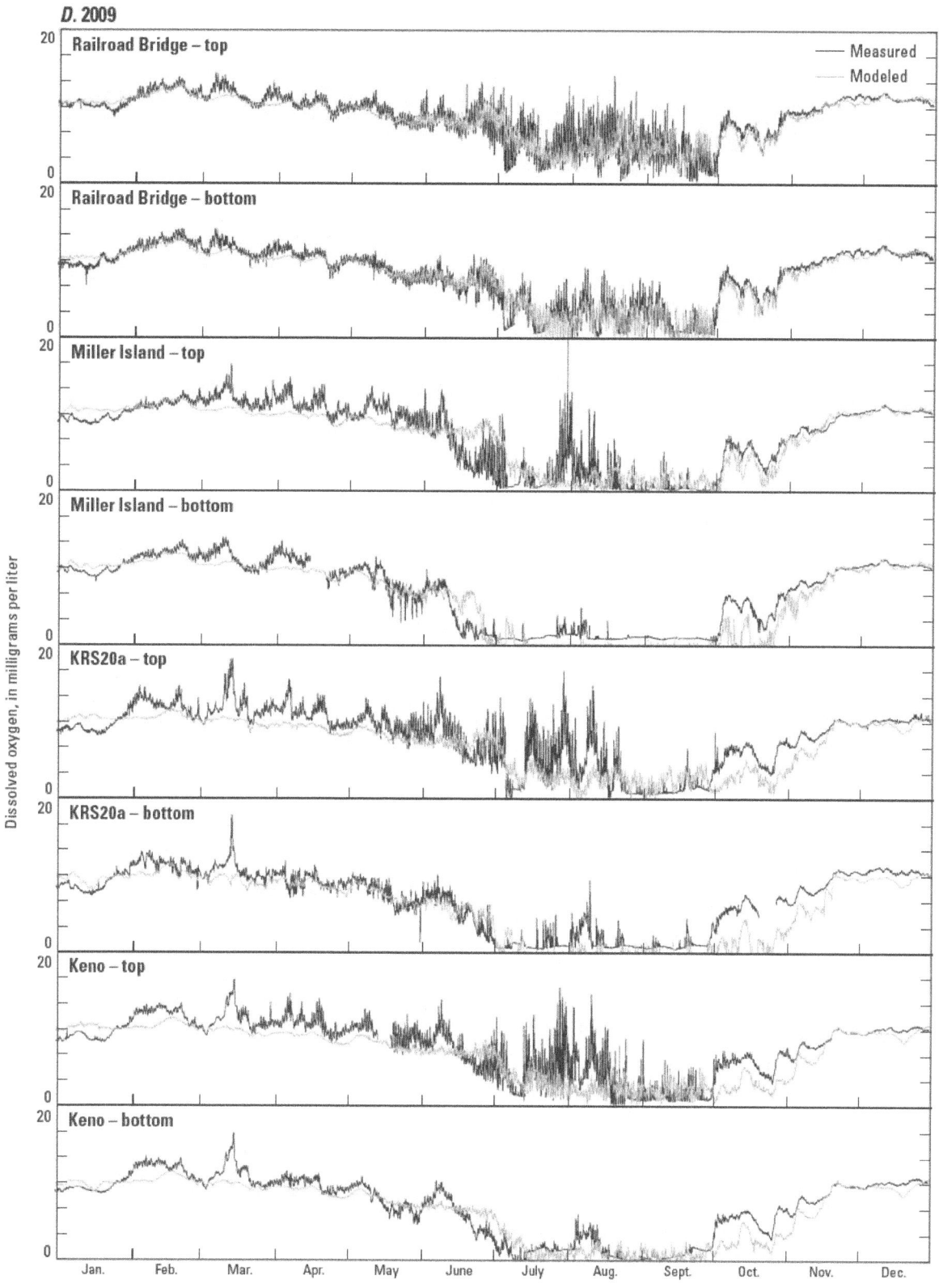

Figure 11.—Continued

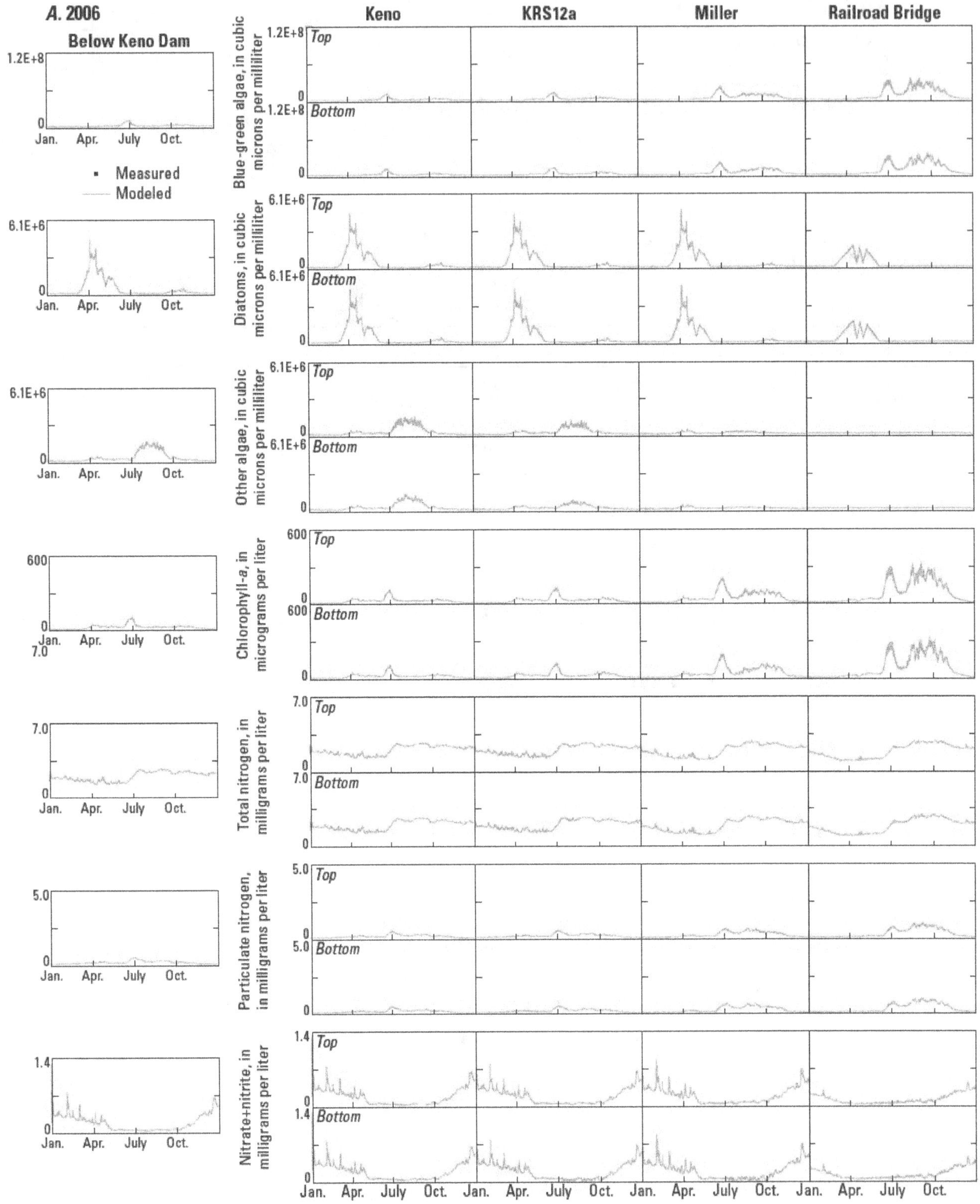

Figure 12. Measured and modeled algae, nutrients, organic matter, and bottom sediment during calendar years (*A*) 2006, (*B*) 2007, (*C*) 2008, and (*D*) 2009 for sites in the Link River to Keno Dam reach of the Klamath River, Oregon. Top results were from 0.5 meters below the river surface; bottom results were from 1 meter above the channel bottom.

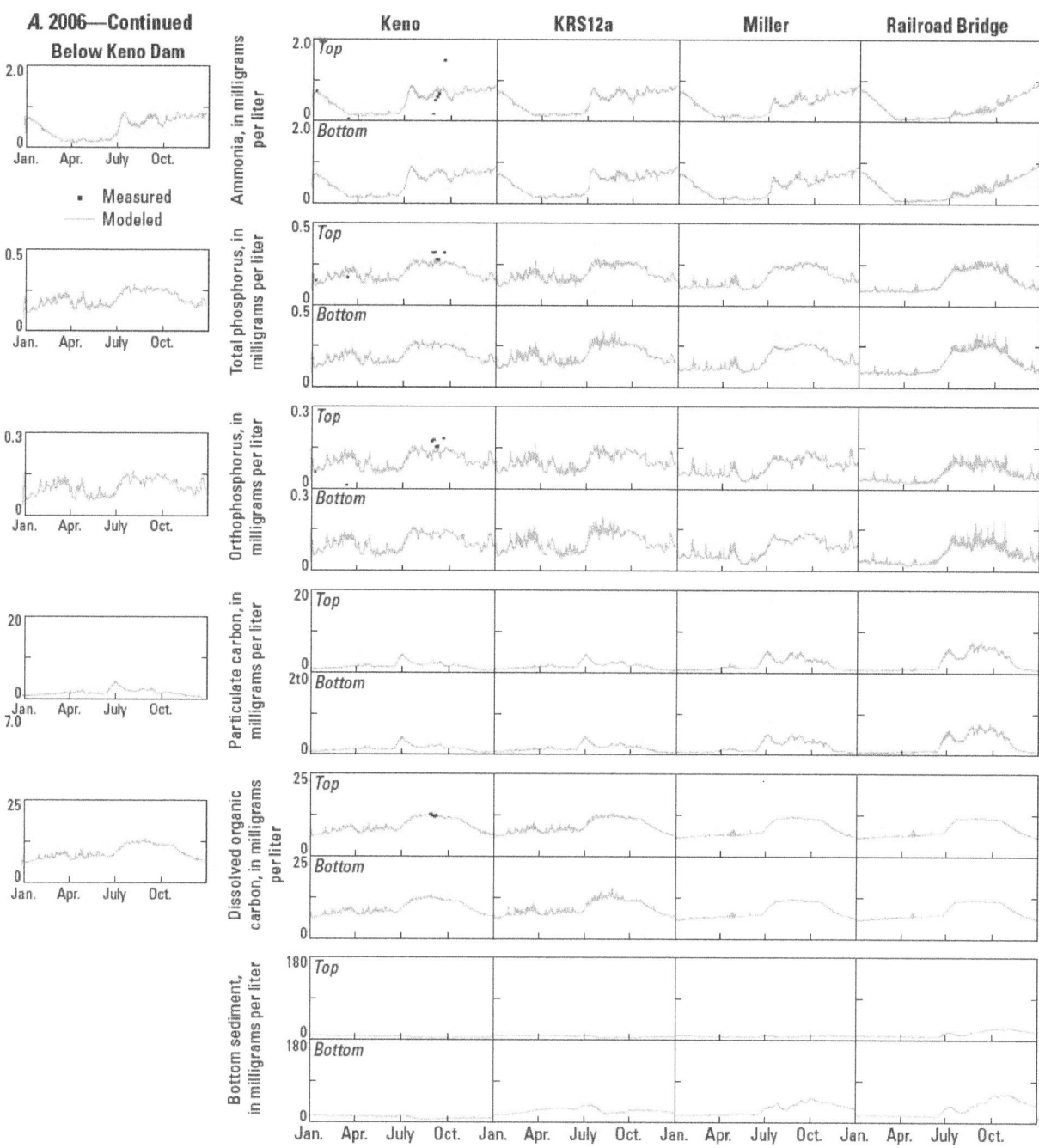

Figure 12.—Continued

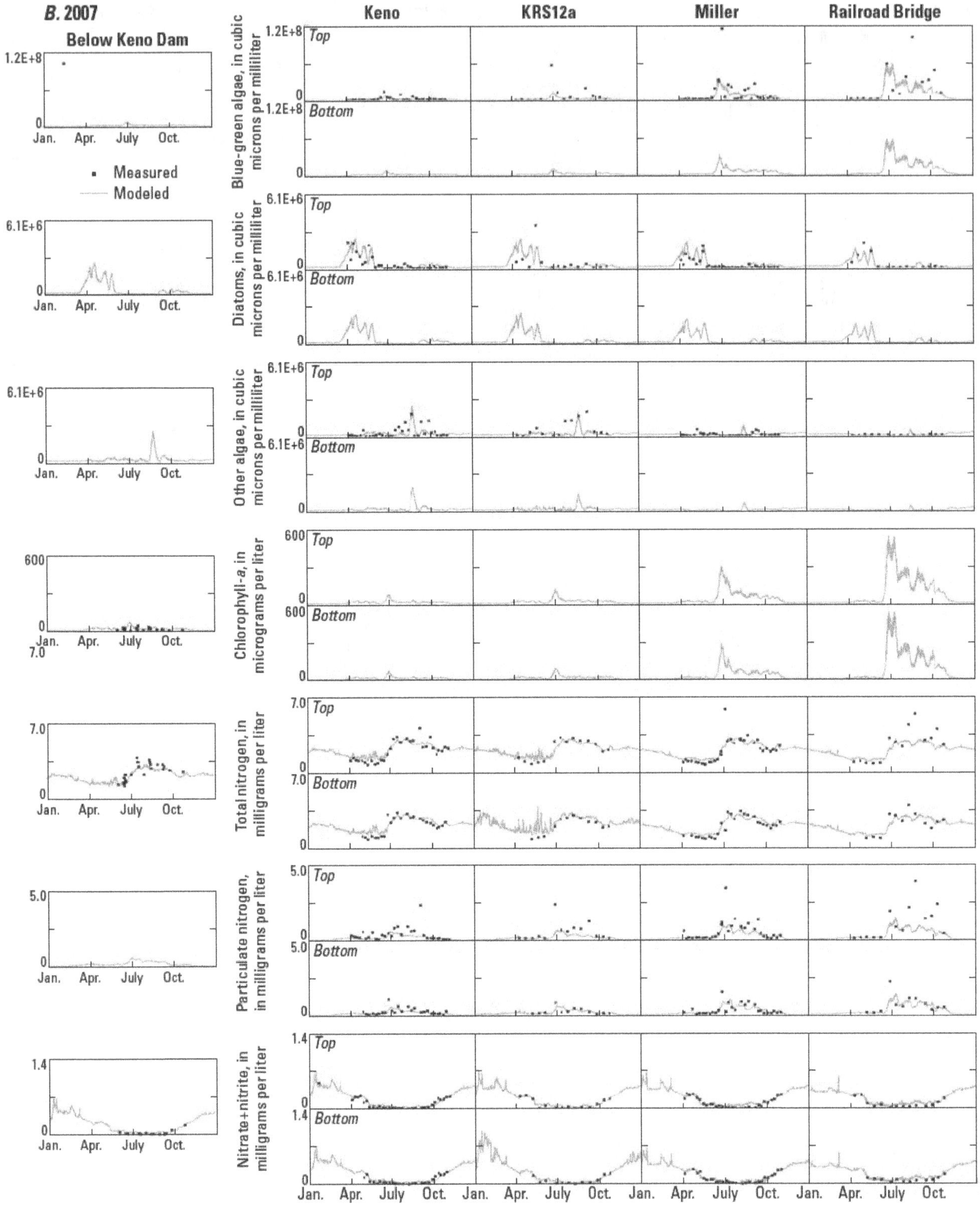

B. 2007

Below Keno Dam

■ Measured
— Modeled

Figure 12.—Continued

B. 2007—Continued

Below Keno Dam

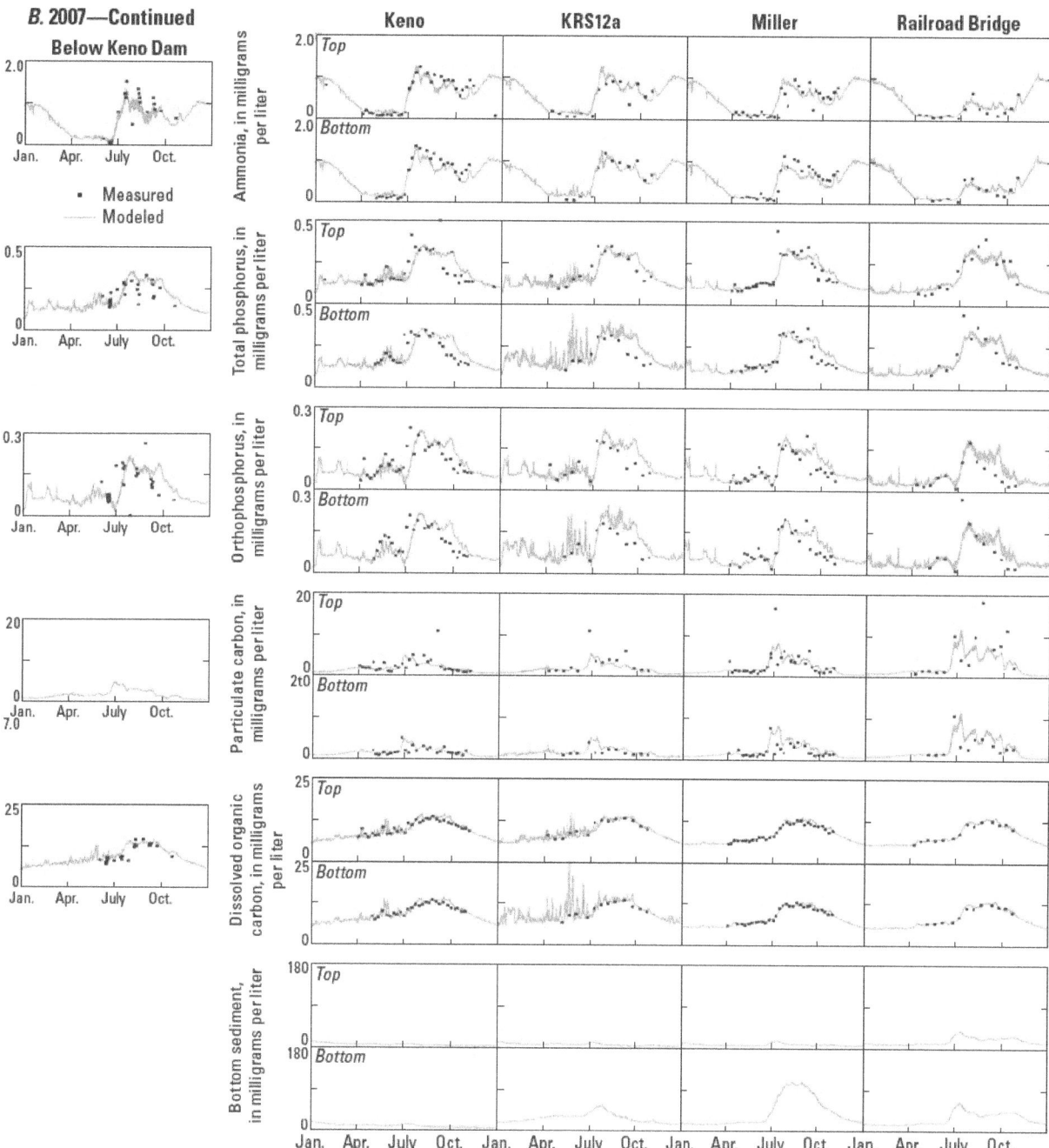

- Measured
— Modeled

Figure 12.—Continued

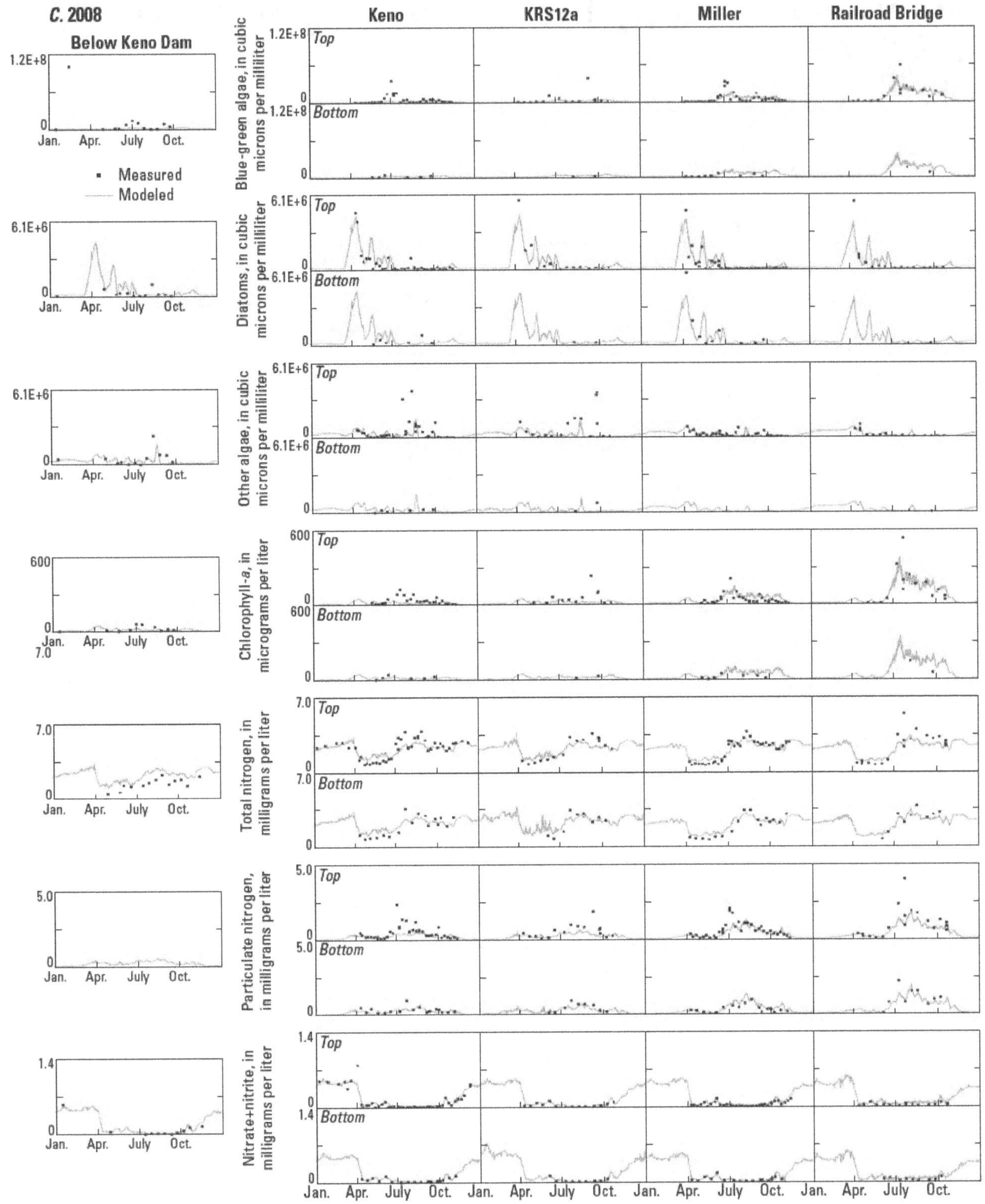

C. 2008

Figure 12.—Continued

C. 2008—Continued

Below Keno Dam

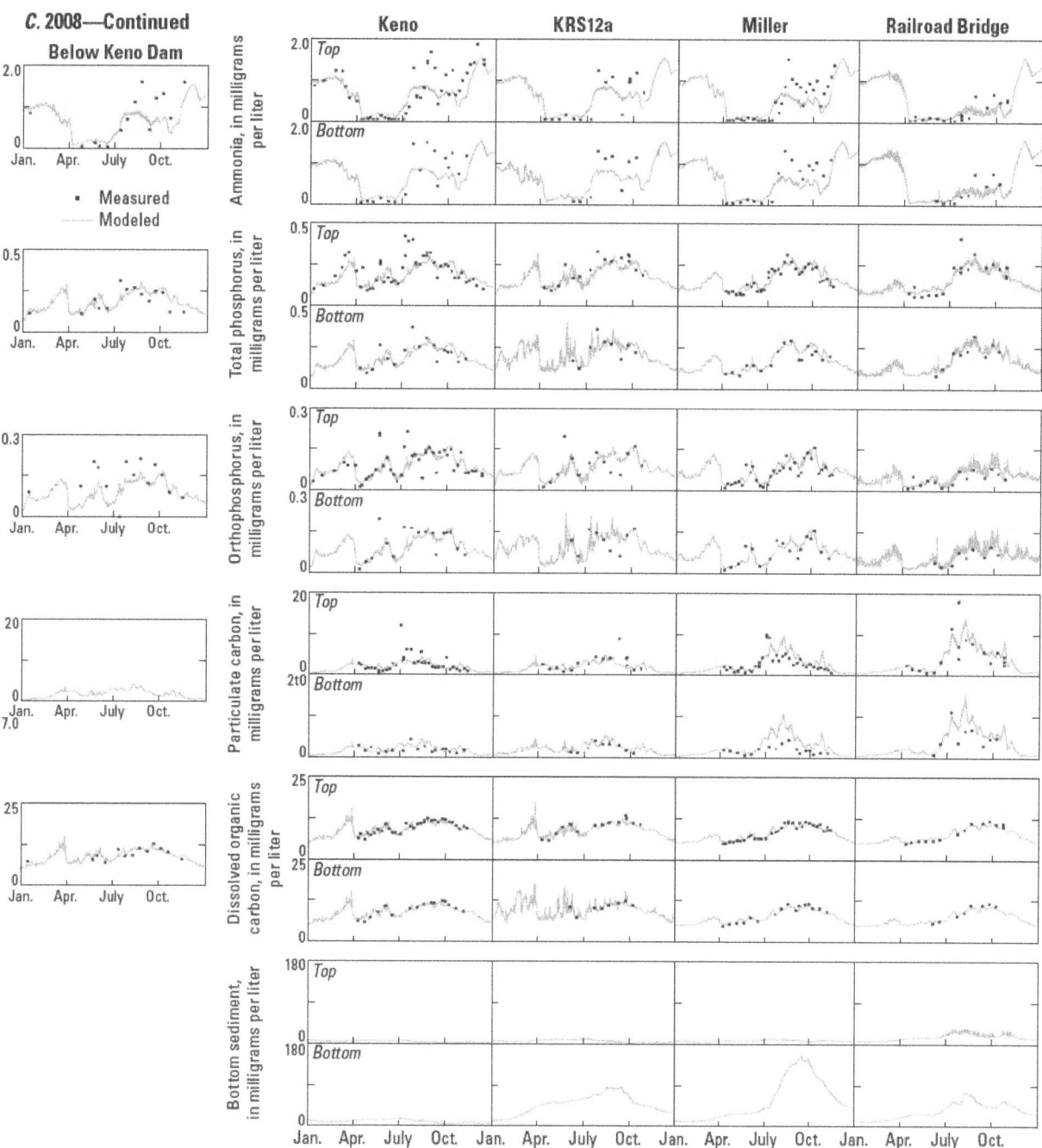

- ▪ Measured
- — Modeled

Figure 12.—Continued

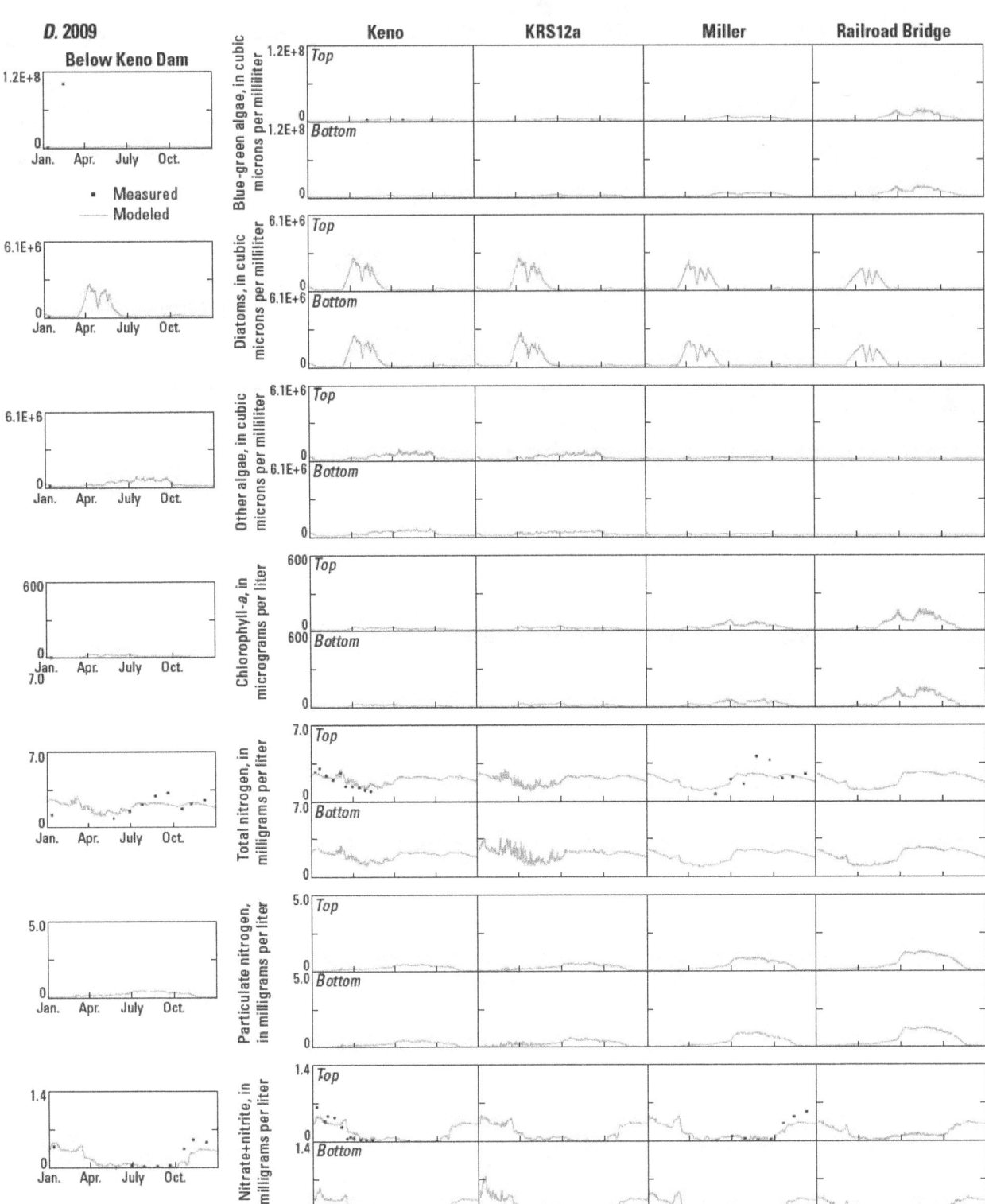

Figure 12.—Continued

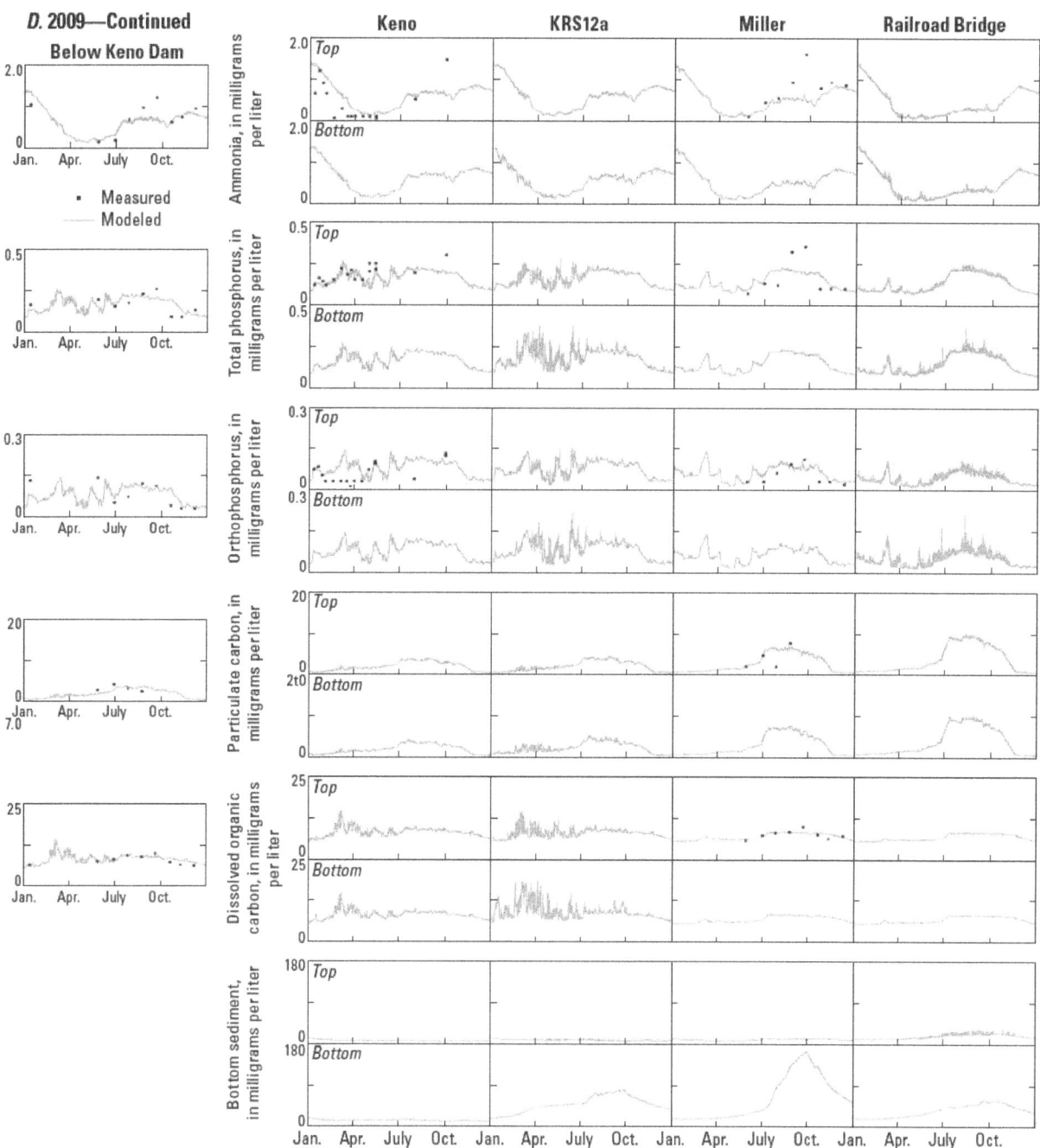

D. 2009—Continued

Below Keno Dam

- Measured
- Modeled

Keno   KRS12a   Miller   Railroad Bridge

Ammonia, in milligrams per liter

Total phosphorus, in milligrams per liter

Orthophosphorus, in milligrams per liter

Particulate carbon, in milligrams per liter

Dissolved organic carbon, in milligrams per liter

Bottom sediment, in milligrams per liter

Figure 12.—Continued

Acknowledgments

Funding for this study was provided by the Bureau of Reclamation. Dean Snyder (U.S. Geological Survey) conducted the alkalinity analyses. Mariah Tilman, Garrett Steensland, Gunter Schanzenbacher, April Tower, and Tracy McGovern (Bureau of Reclamation) assisted with macrophyte field work. Discussions with Rick Carlson and Jason Cameron (Bureau of Reclamation), Steven Gherini (Tetra Tech) and Chris Berger and Scott Wells (Portland State University) were helpful. Thanks to Mark Sytsma, Vanessa Morgan, Rich Miller, and Robyn Draheim (Portland State University Center for Lakes and Reservoirs) for advice on macrophyte sampling and assistance with macrophyte and snail identification.

Supplementary Material

Macrophyte field sampling results are provided in spreadsheet form at: http://or.water.usgs.gov/proj/keno_reach/data.html.

References Cited

Barko, J.W., Adams, M.S., and Clesceri, N.L., 1981, Environmental factors and their consideration in the management of submersed aquatic vegetation: A review: Journal of Aquatic Plant Management, v. 24, p. 1–10.

Barko, J.W. and Smart, R.M., 1981, Comparative influences of light and temperature on the growth and metabolism of selected submersed freshwater macrophytes: Ecological Monographs, v. 51, p. 219–235.

Bates, R.G., 1951, First dissociation constant of phosphoric acid from 0°C to 60°C—Limitations of the electromotive force method for moderately strong acids: Journal of Research of the National Bureau of Standards, v. 47, p. 127–134.

Bates, R.G., and Acree, S.F., 1943, pH values of certain phosphate–chloride mixtures, and the second dissociation constant of phosphoric acid from 0° to 60°C: Journal of Research of the National Bureau of Standards, v. 30, p. 129–155.

Berger, C.J., 2000, Modeling macrophytes of the Columbia Slough: Portland, Oregon, Portland State University, Ph.D. Dissertation, 278 p.

Berger, C.J., and Wells, S.A., 2008, Modeling the effects of macrophytes on hydrodynamics: Journal of Environmental Engineering, v. 134, p. 778–788.

Bonn, B.A., and Fish, William, 1991, Variability in the measurement of humic carboxyl content: Environmental Science and Technology, v. 25, p. 232–240.

Cashatt, D.N., and Bruce, L.J., 2009, Aquatic plant sampling, in Schultz, R.D., ed., Standard gear and techniques for fisheries surveys in Iowa: Des Moines, Iowa, Iowa Department of Natural Resources Completion Report, 186 p.

Chojnacka, K., 2010, Biosorption and bioaccumulation—The prospects for practical applications: Environment International, v. 36, p. 299–307.

Cole, T.M., and Wells, S.A., 2008, CE-QUAL-W2—A two-dimensional, laterally averaged, hydrodynamic and water-quality model, version 3.6: U.S. Army Corps of Engineers, Instruction Report EL-08-1 [variously paged].

Dean, J.A., 1985, Lange's handbook of chemistry (13th ed.): New York, McGraw-Hill, p. 5–16.

Deas, M.L., and Vaughn, J., 2011, Keno Reservoir particulate study 2008—Technical memorandum prepared for the Bureau of Reclamation, Klamath Basin Area Office, April 2011: Davis, California, Watercourse Engineering, Inc., 38 p.

Drever, J.I., 1997, The geochemistry of natural waters—Surface and groundwater environments (3d ed.): New Jersey, Prentice-Hall, 436 p.

Eilers, J.M., 2005, Aquatic vegetation in selected sites of the Lost River, OR and CA: MaxDepth Aquatics, Inc., prepared for Tetra Tech, Inc., January 2005, 18 p.

Emerson, K.R., Russo, C., Lund, R.E., and Thurston, R.V., 1975, Aqueous ammonia equilibrium calculations—Effect of pH and temperature: Journal of the Fisheries Research Board of Canada, v. 32, p. 2379-2383.

Gherini, S.A., Mok, L., Hudson, R.J.M., Davis, G.F., Chen, C.W., and Goldstein, R.A., 1985, The ILWAS model—Formulation and application: Water, Air, and Soil Pollution, v. 26, p. 425–459.

Hauxwell, J., Knight, S., Wagner, K., Mikulyuk, A., Nault, M., Porzky, M., and Chase, S., 2010, Recommended baseline monitoring of aquatic plants in Wisconsin—Sampling design, field and laboratory procedures, data entry and analysis, and applications: Wisconsin Department of Natural Resources Bureau of Science Services PUB-SS 2010, 47 p.

Hem, J.D., 1985, Study and interpretation of the chemical characteristics of natural water (3d ed.): U.S. Geological Survey Water-Supply Paper 2254, p. 105–111. (Also available at http://pubs.usgs.gov/wsp/wsp2254/.)

Hough, R.A., Fornwall, M.D., Negele, B.J., Thompson, R.L., and Putt, D.A., 1989, Plant community dynamics in a chain of lakes—Principal factors in the decline of rooted macrophytes with eutrophication: Hydrobiologia, v. 173, p. 199–217.

Kenow, K.P., Lyon, J.E., Hines, R.K., and Elfessi, A., 2007, Estimating biomass of submersed vegetation using a simple rake sampling technique: Hydrobiologia, v. 575, p. 447–454.

Körner, S., and Nicklisch, A., 2002, Allelopathic growth inhibition of selected phytoplankton species by submerged macrophytes: Journal of Phycology, v. 38, p. 862–871.

Nichols, S.A., and Shaw, B.H., 1986, Ecological life histories of the three aquatic nuisance plants *Myriophyllum spicatum*, *Potamogeton crispus* and *Elodea canadensis*: Hydrobiologica, v. 131, p. 3–21.

Oregon Department of Environmental Quality, 2007, Oregon's 2004/2006 integrated report: Oregon Department of Environmental Quality Web page, accessed November 16, 2007, at http://www.deq.state.or.us/wq/assessment/rpt0406.htm.

Oregon Department of Environmental Quality, 2010, Upper Klamath and Lost River subbasins; total maximum daily load (TMDL) and water quality management plan (WQMP), accessed December 2010, at http://www.deq.state.or.us/WQ/TMDLs/klamath.htm

Owens, C.S., Smart, R.M., Williams, P.E., and Spickard, M.R., 2010, Comparison of three biomass sampling techniques on submersed aquatic plants in a northern tier lake: U.S. Army Corps of Engineers Report ERDC/TN APCRP-EA-24, July 2010, 9 p.

Pankow, J.F., 1991, Aquatic chemistry concepts: Chelsea, Michigan, Lewis Publishers, 673 p.

Perdue, E.M., Reuter, J.H., and Parrish, R.S., 1984, A statistical model of proton binding by humus: Geochimica et Cosmochimica Acta, v. 48, p. 1257–1263.

Perdue, E.M., and Ritchie, J.D., 2003, Dissolved organic matter in freshwaters: Treatise on Geochemistry, v. 5, p. 273–318.

Poulson, S.R., and Sullivan, A.B., 2010, Assessment of diel chemical and isotopic techniques to investigate biogeochemical cycles in the upper Klamath River, Oregon, USA: Chemical Geology, v. 269, p. 3–11.

Press, W.H., Flannery, B.P., Teukolsky, S.A., and Vetterling, W.T., 1989, Numerical recipes—The art of scientific computing (FORTRAN version): Cambridge, Cambridge University Press, 702 p.

Rounds, S.A., 2006, Alkalinity and acid neutralizing capacity (version 4), *in* National field manual for the collection of water-quality data, Wilde, F.D., and Radtke, D.B., eds.: U.S. Geological Survey Techniques of Water-Resources Investigations, book 9, chap. A6, section 6.6, 53 p. (Also available at http://water.usgs.gov/owq/FieldManual/Chapter6/section6.6/.)

Rounds, S.A., and Sullivan, A.B., 2009, Review of Klamath River total maximum daily load models from Link River Dam to Keno Dam, Oregon: U.S. Geological Survey Administrative Report, 37 p. (Also available at http://or.water.usgs.gov/proj/keno_reach/download/klamath_river_model_review_final.pdf.)

Rounds, S.A., and Sullivan, A.B., 2010, Review of revised Klamath River total maximum daily load models from Link River Dam to Keno Dam, Oregon: U.S. Geological Survey Administrative Report, 32 p. (Also available at http://or.water.usgs.gov/proj/keno_reach/download/klamath_model_rereview_final.pdf.)

Stumm, Werner, and Morgan, J.J., 1996, Aquatic chemistry—Chemical equilibria and rates in natural waters (3d ed.): New York, John Wiley and Sons, 1,022 p.

Sullivan, A.B., Deas, M.L., Asbill, J., Kirshtein, J.D., Butler, K., and Vaughn, J., 2009, Klamath River water quality data from Link River Dam to Keno Dam, Oregon, 2008: U.S. Geological Survey Open-File Report 2009–1105, 25 p. (Also available at http://pubs.usgs.gov/of/2009/1105/.)

Sullivan, A.B., Deas, M.L., Asbill, J., Kirshtein, J.D., Butler, K., Stewart, M.A., Wellman, R.E., and Vaughn, J., 2008, Klamath River water quality and acoustic Doppler current profiler data from Link River Dam to Keno Dam, 2007: U.S. Geological Survey Open-File Report 2008–1185, 24 p. (Also available at http://pubs.usgs.gov/of/2008/1185/.)

Sullivan, A.B., Snyder, D.M., and Rounds, S.A., 2010, Controls on biochemical oxygen demand in the upper Klamath River, Oregon: Chemical Geology, v. 269, p. 12–21, doi: 10.1016/j.chemgeo.2009.08.007.

Sullivan, A.B., Rounds, S.A., Deas, M.L., Asbill, J.R., Wellman, R.E., Stewart, M.A., Johnston, M.W., and Sogutlugil, I.E., 2011, Modeling hydrodynamics, water temperature, and water quality in the Klamath River upstream of Keno Dam, Oregon, 2006–09: U.S. Geological Survey Scientific Investigations Report 2011–5105, 70 p. (Also available at http://pubs.usgs.gov/sir/2011/5105.)

Sullivan, A.B., Rounds, S.A., Deas, M.L., and Sogutlugil, I.E., 2012, Dissolved oxygen analysis, TMDL model comparison, and particulate matter shunting—Preliminary results from three model scenarios for the Klamath River upstream of Keno Dam, Oregon: U.S. Geological Survey Open-File Report 2012–1101, 30 p. (Also available at http://pubs.usgs.gov/of/2012/1101.)

Tetra Tech, Inc., 2009, Klamath River model for TMDL development, Prepared for U.S. Environmental Agency Region 9 and 10, Oregon Department of Environmental Quality, and North Coast Regional Water Quality Control Board, December 2009: 196 p., accessed May 20, 2011, at http://www.deq.state.or.us/wq/tmdls/docs/klamathbasin/uklost/KlamathLostAppendixC.pdf.

Tipping, E., 1994, WHAM—A chemical equilibrium model and computer code for waters, sediments, and soils incorporating a discrete site/electrostatic model of ion-binding by humic substances: Computers and Geosciences, v. 20, p. 973–1023.

Appendix A. CE-QUAL-W2 Macrophyte Code Changes

The most important macrophyte code changes related to initial conditions are included here. Other minor code changes also were made, including an expansion of macrophyte model flux outputs. For the definitive list of changes, users should compare new and old source code files.

Normal macrophyte initialization is invoked by setting MACWBCI to a value ≥ 0. With the new code changes, however, if MACWBCI is set to a value ≤ -1.0, a longitudinal set of initial concentrations is read from the longitudinal profile input file for every cell in the grid. The cell concentrations are read in units of grams per cubic meter, the same units used for MACWBCI. The model reads initial concentrations for each macrophyte group separately and in order, with one line per segment, and the initial concentrations for each layer read from KT to KB (left to right) as 8-character floating point inputs. This is the same input format as other longitudinal inputs; the macrophyte initial concentrations are placed at the end of that input file.

New variables include LONG_MACROPHYTE() and MACCI(). LONG_MACROPHYTE() is a logical, "true" when the user asks to read initial macrophyte concentrations from a longitudinal profile file. MACCI() is a real array holding the longitudinal initial concentrations. The longitudinal concentrations are read and initial macrophyte concentrations and masses are set in the init-cond f90 source-code file:

```
DO I=cus(jb),ds(jb)
  IF (LONG_MACROPHYTE(JW,M)) READ (LPR(JW),'(//(8X,9F8.0))') (MACCI(K,I,M),K=KT,KB(I))    ! SR 12/21/11

  depkti=ELWS(i)-el(kti(i)+1,i)
  if(depkti.ge.thrkti)then
    kticol(i)=.true.
    jt=kti(i)
  else
    kticol(i)=.false.
    jt=kti(i)+1
  end if

  je=kb(i)
  DO j=jt,je
    if(j.le.kt)then
      k=kt
    else
      k=j
    end if
    IF (LONG_MACROPHYTE(JW,M)) THEN               ! SR 12/21/11
      MACRC(J,K,I,M) = MACCI(K,I,M)               ! SR 12/21/11
    ELSE                                          ! SR 12/21/11
      macrc(j,K,I,m) = macwbci(jw,m)
    END IF                                        ! SR 12/21/11
    SMACRC(J,K,I,M) = MACRC(J,K,I,M)              ! SR 12/21/11
  END DO
END DO
```

Macrophyte initializations are also dealt with in the layeraddsub.f90 and w2_36_gen.f90 source files.

Appendix B. CE-QUAL-W2 pH and Alkalinity Code Changes

The most important code changes made to the CE-QUAL-W2 pH and alkalinity routines are included in this section. For a definitive list of all changes, those who are interested should compare the new and old source files electronically.

The enhanced pH buffering routines and the nonconservative alkalinity algorithms can be turned on or off by the user through two new global input parameters in the control file on the MISCELL card; that modified card is shown below:

```
MISCELL    NDAY  PHBUFC  NCALKC
           100     ON      ON
```

The PHBUFC variable (ON/OFF) controls the use of the enhanced pH buffering routines; the NCALKC variable (ON/OFF) controls the use of the nonconservative alkalinity algorithms.

If the enhanced pH buffering routines are turned on, then the model will read a new input file named "ph_buffering npt" that takes the following form:

```
Enhanced pH Buffering Input File for CE-QUAL-W2

BUFTYPE NH4BUFC PO4BUFC OMBUFC
           ON      ON      ON


OM TYPE   OMTYPE    NAG POMBUFC
           DIST      2    OFF


DENSITY    SDEN    SDEN   SDEN   SDEN   SDEN   SDEN   SDEN   SDEN   SDEN
           0.14    0.10


pK VALS      PK      PK     PK     PK     PK     PK     PK     PK     PK
            4.5     9.6


STD DEV    PKSD    PDSD   PDSD   PDSD   PDSD   PDSD   PDSD   PDSD   PDSD
            1.2     1.0


----------------------------------------------------------------------------
Input Variables:
  NH4BUFC  ON/OFF, specifies whether ammonia/ammonium is included in pH buffering
  PO4BUFC  ON/OFF, specifies whether phosphoric acid is included in pH buffering
  OMBUFC   ON/OFF, specifies whether organic matter is included in pH buffering
  OMTYPE   DIST or MONO
           where DIST specifies one or more Gaussian distributions of pKa values,
              or MONO specifies a collection of discrete pKa values
  NAG      the number of acid/base groups to model, either as the means of
           Gaussian distributions of pKa values or as discrete monoprotic acids
  POMBUFC  ON/OFF, specifies whether POM is included in OM buffering
           where ON indicates that OM buffering includes both DOM and POM
               OFF indicates that OM buffering includes only DOM
  SDEN     site density, in moles of acid/base sites per mole of carbon in OM
  PK       the pKa values (negative log10 of the acid dissociation constant),
           specified either as the mean of a distribution or a discrete value
  PKSD     the standard deviation for a Gaussian distribution of pKa values
           (ignored when specifying an OMTYPE of MONO)
```

When the model is run with the enhanced pH buffering capability turned on, the code will generate a new output file named "ph_buffering.opt" based on the new inputs from the ph_buffering npt file. The object is to echo some of the inputs and provide the user with information on the distribution of site densities and pK$_a$ values for the organic matter acids. The following is the output that corresponds to the input file shown above:

```
Enhanced pH buffering output file
Ammonia buffering:   ON
Phosphate buffering: ON
OM buffering:        ON
POM buffering:       OFF
OM buffer type:      DIST

Inputs:
Group  Site density   pKa    std.dev.
  1        0.1400     4.500    1.200
  2        0.1000     9.600    1.000

Modeled:
Group  Site density   pKa
  1        0.0001     0.500
  2        0.0003     1.000
  3        0.0010     1.500
  4        0.0027     2.000
  5        0.0058     2.500
  6        0.0107     3.000
  7        0.0164     3.500
  8        0.0213     4.000
  9        0.0233     4.500
 10        0.0213     5.000
 11        0.0165     5.500
 12        0.0107     6.000
 13        0.0060     6.500
 14        0.0033     7.000
 15        0.0032     7.500
 16        0.0059     8.000
 17        0.0110     8.500
 18        0.0167     9.000
 19        0.0199     9.500
 20        0.0184    10.000
 21        0.0133    10.500
 22        0.0075    11.000
 23        0.0033    11.500
 24        0.0011    12.000
 25        0.0003    12.500
 26        0.0001    13.000
 27        0.0000    13.500
```

The "ph_buffering npt" input file is read by using new code in the input.f90 source file, including the following:

```
! Initialize variables for enhanced pH buffering          ! entire section  ! SR 01/01/12

IF (CONSTITUENTS .AND. PHBUFC == '      ON') THEN
  OPEN (298,FILE='ph_buffering.npt',STATUS='OLD')
  READ (298,'(///8X,3A8)')     NH4BUFC, PO4BUFC, OMBUFC
  READ (298,'(//8X,A8,I8,A8)') OMTYPE,  NAGI,    POMBUFC

  ALLOCATE (SDENI(NAGI),PKI(NAGI),PKSD(NAGI))

  READ (298,'(//(:8X,9F8.0))') (SDENI(J), J=1,NAGI)
  READ (298,'(//(:8X,9F8.0))') (PKI(J),   J=1,NAGI)
  READ (298,'(//(:8X,9F8.0))') (PKSD(J),  J=1,NAGI)
  CLOSE (298)

  AMMONIA_BUFFERING   = NH4BUFC == '      ON'
  PHOSPHATE_BUFFERING = PO4BUFC == '      ON'
  OM_BUFFERING        = OMBUFC  == '      ON'
  POM_BUFFERING       = POMBUFC == '      ON' .AND. OM_BUFFERING

  IF (OM_BUFFERING) THEN
    SDENI = ABS(SDENI)
    IF (OMTYPE == '    DIST') THEN
      IF (ANY(PKSD <= 0)) THEN
        WARNING_OPEN = .TRUE.
        WRITE (WRN,'(A)')  'WARNING -- PKSD inputs in the ph_buffering.npt file must be greater than zero.'
        WRITE (WRN,'(A/)') 'Please fix your inputs. For now, PKSD values of zero will be set to 1.'
      END IF
      DO JA=1,NAGI
        IF (PKSD(JA) <= 0) PKSD(JA) = 1.0
      END DO

      NAG = 27
      ALLOCATE (SDEN(NAG),PK(NAG),FRACT(NAG))
      SDEN = 0.0
      DO J=1,NAG
        PK(J) = 0.5*J
      END DO
      DO JA=1,NAGI
        SUM = 0.0
        DO J=1,NAG
          FRACT(J) = EXP(-0.5*(((PK(J)-PKI(JA))/PKSD(JA))**2))
          SUM      = SUM+FRACT(J)
        END DO
        DO J=1,NAG
          SDEN(J) = SDEN(J)+SDENI(JA)*FRACT(J)/SUM
        END DO
      END DO
```

```
      ELSE
        ALLOCATE (SDEN(NAGI),PK(NAGI))
         NAG    = NAGI
         SDEN   = SDENI
         PK     = PKI
         OMTYPE = '    MONO'
      END IF
   END IF


   OPEN (299,FILE='ph_buffering.opt',STATUS='UNKNOWN')
   WRITE (299,'(A/)') 'Enhanced pH buffering output file'
   WRITE (299,'(2A)') 'Ammonia buffering:   ', ADJUSTL(TRIM(NH4BUFC))
   WRITE (299,'(2A)') 'Phosphate buffering: ', ADJUSTL(TRIM(PO4BUFC))
   WRITE (299,'(2A)') 'OM buffering:        ', ADJUSTL(TRIM(OMBUFC))
   IF (OM_BUFFERING) THEN
     WRITE (299,'(2A)') 'POM buffering:       ', ADJUSTL(TRIM(POMBUFC))
     WRITE (299,'(2A)') 'OM buffer type:      ', ADJUSTL(TRIM(OMTYPE))
     WRITE (299,'(/A/A)') 'Inputs:','Group  Site density   pKa    std.dev.'
     DO JA=1,NAGI
       IF (OMTYPE == '    DIST') THEN
         WRITE (299,'(1X,I3,5X,F8.4,4X,F6.3,3X,F6.3)') JA, SDENI(JA), PKI(JA), PKSD(JA)
       ELSE
         WRITE (299,'(1X,I3,5X,F8.4,4X,F6.3,3X,A)') JA, SDENI(JA), PKI(JA), ' N/A'
       END IF
     END DO
     WRITE (299,'(/A/A)') 'Modeled:','Group  Site density   pKa'
     DO JA=1,NAG
       WRITE (299,'(1X,I3,5X,F8.4,4X,F6.3)') JA, SDEN(JA), PK(JA)
     END DO
   END IF
   CLOSE (299)


ELSE
  AMMONIA_BUFFERING   = .FALSE.
  PHOSPHATE_BUFFERING = .FALSE.
  OM_BUFFERING        = .FALSE.
  POM_BUFFERING       = .FALSE.
END IF
```

In the water-quality f90 source file, the PH_CO2 subroutine was modified to include the optional buffering by ammonia, phosphoric acid, and organic matter. The modified code is:

```
ENTRY PH_CO2                              ! Enhancements added for buffering by ammonia, phosphate, and OM   ! SR 01/01/12

! pH and carbonate species

  DO I=IU,ID
    DO K=KT,KB(I)
      T1K   = T1(K,I)+273.15
      CART  = TIC(K,I)/12011.                                                                                ! SR 01/01/12
      ALKT  = ALK(K,I)/50044.                                                                                ! SR 01/01/12
      AMMT  = NH4(K,I)/14006.74                                                                              ! SR 01/01/12
      PHOST = PO4(K,I)/30973.762                                                                             ! SR 01/01/12
      OMCT  = (LDOM(K,I)+RDOM(K,I))*ORGC(JW)/12011.           ! moles carbon per liter from DOM             ! SR 01/01/12
      IF (POM_BUFFERING) OMCT = OMCT + (LPOM(K,I)+RPOM(K,I))*ORGC(JW)/12011.                                 ! SR 01/01/12


!**** Ionic strength

      IF (FRESH_WATER(JW)) S2 = 2.5E-05*TDS(K,I)
      IF (SALT_WATER(JW))  S2 = 1.47E-3+1.9885E-2*TDS(K,I)+3.8E-5*TDS(K,I)*TDS(K,I)


!**** Debye-Huckel terms and activity coefficients

      SQRS2  = SQRT(S2)
      DH1    = -0.5085*SQRS2/(1.0+1.3124*SQRS2)+4.745694E-03+4.160762E-02*S2-9.284843E-03*S2*S2
      DH2    = -2.0340*SQRS2/(1.0+1.4765*SQRS2)+1.205665E-02+9.715745E-02*S2-2.067746E-02*S2*S2
      DH3    = -4.5765*SQRS2/(1.0+1.3124*SQRS2)                          ! extended Debye-Huckel for PO4     ! SR 01/01/12
      DHH    = -0.5085*SQRS2/(1.0+2.9529*SQRS2)                          ! extended Debye-Huckel for H+ ion  ! SR 01/01/12
      H2CO3T = 10.0**(0.0755*S2)
      HCO3T  = 10.0**DH1
      CO3T   = 10.0**DH2
      PO4T   = 10.0**DH3                                                                                     ! SR 01/01/12
      HT     = 10.0**DHH         ! activity coefficient for H+                                               ! SR 01/01/12
      HPO4T  = CO3T              ! tabled values similar to those for carbonate                              ! SR 01/01/12
      OHT    = HCO3T             ! tabled values similar to those for bicarbonate                            ! SR 01/01/12
      H2PO4T = HCO3T             ! tabled values similar to those for bicarbonate                            ! SR 01/01/12
      NH4T   = HCO3T             ! tabled values similar to those for bicarbonate                            ! SR 01/01/12
      NH3T   = H2CO3T            ! neutral species, set coefficient to same as that for carbonic acid        ! SR 01/01/12
      H3PO4T = H2CO3T            ! neutral species, set coefficient to same as that for carbonic acid        ! SR 01/01/12


!**** Temperature adjustment

      KW   = 10.0**(-283.971  -0.05069842*T1K +13323.0/T1K  +102.24447*LOG10(T1K) -1119669.0/(T1K*T1K))/OHT
      K1   = 10.0**(-356.3094 -0.06091964*T1K +21834.37/T1K +126.8339 *LOG10(T1K) -1684915  /(T1K*T1K))*H2CO3T/HCO3T
      K2   = 10.0**(-107.8871 -0.03252849*T1K + 5151.79/T1K + 38.92561*LOG10(T1K) - 563713.9/(T1K*T1K))*HCO3T/CO3T
      KAMM = 10.0**(-0.09018 -2729.92/T1K)*NH4T/NH3T                                                         ! SR 01/01/12
      KP1  = 10.0**(4.5535 -0.013486*T1K -799.31/T1K)*H3PO4T/H2PO4T     ! Bates (1951)                       ! SR 01/21/12
      KP2  = 10.0**(5.3541 -0.019840*T1K -1979.5/T1K)*H2PO4T/HPO4T      ! Bates and Acree (1943)             ! SR 01/21/12
      KP3  = 10.0**(-12.38) *HPO4T/PO4T                                 ! Dean (1985)                        ! SR 01/01/12
```

```
!**** pH evaluation

        PHT = -PH(K,I)-2.1
        IF (PH(K,I) <= 0.0) PHT = -14.0
        INCR = 10.0
        DO N=1,3
          F    = 1.0
          INCR = INCR/10.0
          ITER = 0
          DO WHILE (F > 0.0 .AND. ITER < 12)
            PHT  = PHT+INCR
            HION = 10.0**PHT
            F    = CART*K1*(HION+2.0*K2)/(HION*HION+K1*HION+K1*K2)+KW/HION-ALKT-HION/HT      ! SR 01/01/12
            IF (AMMONIA_BUFFERING) THEN                                                     ! SR 01/01/12
              F  = F + AMMT*KAMM/(HION+KAMM)                                                ! SR 01/01/12
            END IF                                                                          ! SR 01/01/12
            IF (PHOSPHATE_BUFFERING) THEN                                                   ! SR 01/01/12
              F  = F + PHOST*( KP1*KP2*HION + 2*KP1*KP2*KP3 - HION*HION*HION )                          &
                           /( HION*HION*HION + KP1*HION*HION + KP1*KP2*HION + KP1*KP2*KP3)   ! SR 01/01/12
            END IF                                                                          ! SR 01/01/12
            IF (OM_BUFFERING) THEN                                                          ! SR 01/01/12
              DO JA=1,NAG                                                                   ! SR 01/01/12
                F = F + OMCT*SDEN(JA)*( 1.0/(1.0+HION*(10.0**PK(JA))) - 1.0/(1.0+(10.0**(PK(JA)-4.5))) ) ! SR 01/01/12
              END DO                                                                        ! SR 01/01/12
            END IF                                                                          ! SR 01/01/12
            ITER = ITER+1
          END DO
          PHT = PHT-INCR
        END DO

!**** pH, carbon dioxide, bicarbonate, and carbonate concentrations

        HION      = 10.0**PHT
        PH(K,I)   = -PHT
        CO2(K,I)  = TIC(K,I)/(1.0+K1/HION+K1*K2/(HION*HION))
        HCO3(K,I) = TIC(K,I)/(1.0+HION/K1+K2/HION)
        CO3(K,I)  = TIC(K,I)/((HION*HION)/(K1*K2)+HION/K2+1.0)
      END DO
    END DO
  RETURN
```

In the water-quality.f90 source file, a new ALKALINITY subroutine was added to realize the nonconservative components of alkalinity. The new code is:

```
!*****************************************************************************************************
!**                                     A L K A L I N I T Y                                        **
!*****************************************************************************************************

ENTRY ALKALINITY                                            ! entire subroutine added   ! SR 01/01/12

! According to Stumm and Morgan (1996), table 4.5 on page 173:
! Utilization of ammonium during photosynthesis results in an alkalinity decrease:  14 eq. alk per 16 moles ammonium
! Utilization of nitrate during photosynthesis results in an alkalinity increase:  18 eq. alk per 16 moles nitrate
! Production of ammonium during respiration results in an alkalinity increase:   14 eq. alk per 16 moles ammonium
! Nitrification of ammonium results in an alkalinity decrease:                    2 eq. alk per  1 mole  ammonium
! Denitrification of nitrate (to nitrogen gas) results in an alkalinity increase:  1 eq. alk per  1 mole  nitrate

! Alkalinity is represented as mg/L CaCO3 (MW=100.088).  CaCO3 has 2 equivalents of alk per mole.
! Nitrogen has an atomic mass of 14.00674.  These numbers account for the factor of 50.044/14.00674 used below.

  DO I=IU,ID
    DO K=KT,KB(I)
      ALKSS(K,I) = (50.044/14.00674) * ( 14./16.*(NH4AP(K,I)+NH4EP(K,I)+NH4ZR(K,I)+NH4MR(K,I)-NH4MG(K,I))        &
                                       + 18./16.*(NO3AG(K,I)+NO3EG(K,I))                                         &
                                       - 2.*NH4D(K,I) + NO3D(K,I) + NO3SED(K,I)*(1-FNO3SED(JW)) )
    END DO
  END DO
RETURN
```

www.ingramcontent.com/pod-product-compliance
Lightning Source LLC
Chambersburg PA
CBHW081614170526
45166CB00009B/2965